JUL 0 2

Lifting Titan's Veil

Lifting Titan's Veil is a revealing account of the second largest moon in
our solar system. This world in orbit around Saturn is the only body in the
solar system with an atmosphere strikingly similar to Earth's and the
only moon with a substantial atmosphere. Nitrogen is the main gas in
Titan's atmosphere but it is laced with a cocktail of hydrocarbons and is
virtually opaque to human eyes because of layers of orange smog.
Beneath the haze, lakes of liquid methane may be a feature of the frigid
landscape. Titan is like a giant laboratory in deep freeze that may help
scientists understand the first chemical steps towards the origin of life.

Beginning with its discovery in 1655, the authors describe our current
knowledge of Titan, including observations made before the space age,
results from the exciting *Voyager* missions of the 1980s, and the most
recent revelations from the world's most advanced telescopes. Ralph
Lorenz has been closely involved with the *Cassini* mission, which will
reach Saturn in 2004 and release the *Huygens* probe into Titan's atmosphere in 2005. Looking forward with anticipation to the new discoveries
Cassini–Huygens is expected to make, he includes some of his personal
experiences in preparing for the mission. Jacqueline Mitton brings a
wealth of experience in writing accessible books on astronomy.

This book will appeal to readers interested in astronomical discovery
and space exploration. It is a splendid introduction to Titan, and essential
reading for anyone who wants to be ready for the arrival of
Cassini–Huygens in 2004.

RALPH LORENZ trained as an engineer and worked for the European
Space Agency at the very beginning of the *Huygens* project. Since obtaining a Ph.D. at the University of Kent, England, he has worked as a planetary scientist at the University of Arizona, USA. His research interests
focus on Titan, but also include climatology, radar, impact dynamics and
spacecraft and instrumentation design. He has been involved in NASA's
largest planetary mission (*Cassini*) and its smallest (the *DS-2 Mars
Microprobes*).

JACQUELINE MITTON obtained a Ph.D. in astrophysics from the
University of Cambridge, and is now a full-time writer and media consul-
tant specialising in astronomy. She has acted as Press Officer for the Royal
Astronomical Society since 1989, and was Editor of the *Journal of the
British Astronomical Association* 1989–1993. She has written or co-
authored sixteen published astronomy books.

Lifting Titan's Veil

Exploring the giant moon of Saturn

Ralph Lorenz
and Jacqueline Mitton

CAMBRIDGE
UNIVERSITY PRESS

PUBLISHED BY THE PRESS SYNDICATE OF THE UNIVERSITY OF CAMBRIDGE
The Pitt Building, Trumpington Street, Cambridge, United Kingdom

CAMBRIDGE UNIVERSITY PRESS
The Edinburgh Building, Cambridge CB2 2RU, UK
40 West 20th Street, New York, NY 10011–4211, USA
477 Williamstown Road, Port Melbourne, VIC 3207, Australia
Ruiz de Alarcón 13, 28014 Madrid, Spain
Dock House, The Waterfront, Cape Town 8001, South Africa

http://www.cambridge.org

First published 2002

Printed in the United Kingdom at the University Press, Cambridge

Typeface Nimrod MT 9/14 pt   *System* QuarkXPress™   [SE]

A catalogue record for this book is available from the British Library

ISBN 0 521 79348 3 hardback

Contents

Colour plate section between pages 152–153

Preface

This book is about several things. Our knowledge of Titan has expanded enormously in the last decade and, first and foremost, we have written about Titan itself, a mysterious world of ice and rock orbiting Saturn. But science is not a smooth and steady ascent to enlightenment. There are false steps and stumbles along the way. The story of *how* we have found out about Titan is as fascinating as *what* we actually know. This is our second major theme. Finally, we anticipate the arrival at Saturn of the *Cassini–Huygens* mission, tracing the development of this massive international project and speculating on what it might find when it reaches its destination.

Planetary exploration is a huge collaborative undertaking, and it has not been possible to mention every participant by name. Similarly, many people, organisations and parts of the *Cassini* project have not been included. Since this particular book is about Titan, we have included the features and details of the mission most relevant to our topic.

Much of the story we tell here is written from personal experience but, to avoid a schizophrenic style, the particularly personal episodes are distinguished by passages under the heading 'Ralph's Log'. The reader could skip these without losing much of the thread of the book, although we'd urge that you do not do so – we think the personal insights here add a unique dimension to our account. (For the record, no such 'Log' exists – the episodes and details have been recorded here from memory. Lorenz accepts responsibility for any inaccuracies.) There is of course a tendency in describing investigations in which one is involved to emphasise ones own contributions. This book is probably not immune from such tendencies even though we have endeavoured to give a balanced account overall.

Acknowledgements

We are indebted to our many colleagues who answered questions, and directed us to reference material and sources of illustrations. Others gave permission to use graphic material, for which we are grateful, including Gordon Garrard, Tetsuya Tokano and Mark Garlick. Special thanks are due to James Garry. A critical reading of the manuscript by Zibi Turtle improved it considerably.

Last but not least, we thank the American and European taxpayers, without whom there would be few discoveries to write about or look forward to.

Ralph Lorenz
Tucson, Arizona, USA

Jacqueline Mitton
Cambridge, UK

April 2001

Discovering Titan

The landscape seems alien. The few clouds that burned a garish red as the Sun set have flitted away and the sky is clear. Strange and unfamiliar life-forms, deprived of water, struggle to survive in the harsh conditions. This is no extraterrestrial scene though, but Tucson, Arizona. Arizona's commendably dark and clear skies are a magnetic attraction for astronomers.

High above, Jupiter gleams brilliantly. Through even a small telescope, an entourage of four moons circling this giant planet and its cloud bands alternating light and dark are obvious. A short distance to the east there is another planet, not as bright as Jupiter. It's Saturn. Through the telescope it is an altogether different object, with its rings tilted tastefully – as though a jeweller had set it there. A little to one side of the rings is a dim, unprepossessing dot, looking a little reddish maybe. This dot is the focus of our attention – Saturn's moon, Titan, a world as intriguing as any in the solar system.

On the 15th of October 1997, another Titan roared into the sky. To be precise, it was a Titan IVB/Centaur launch vehicle. Just before 5 o'clock in the morning local time the appropriately named rocket blasted off from Cape Canaveral Air Force Station, Florida, bearing a 5.8-tonne spacecraft bound for Saturn and Titan. It was the start of a seven-year journey for the *Cassini–Huygens* mission and of a tantalising seven-year wait for the anxious scientists on the ground. *Cassini* was destined to enter orbit around Saturn on the 1st of July 2004. Seven months later, if all goes according to plan, the *Huygens* probe will detach itself, cruise towards Titan for about three weeks, then parachute down onto Titan's surface. Instruments on board the *Cassini* orbiter will gather data about Saturn and its moons, espe-

Figure 1.1. The launch of the *Cassini–Huygens* mission on the 15th of October 1997 at 4.43 a.m. EDT, from Cape Canaveral Air Station in Florida. The launch vehicle was a Titan IVB/Centaur. NASA image. (In colour as Plate 1.)

cially Titan, over a four-year period. Then Titan will become the most distant world by far to have a human artefact land upon it. The enormous effort dedicated to achieving this feat is a testament to the growth in our fascination with Titan as world of unique significance in the quest to understand our own planet.

To see how Titan became the centre of such attention we must first turn the calendar back to the middle of the seventeenth century.

Galileo and the Saturn enigma

When Galileo Galilei turned a crude, low-powered telescope to the sky he opened a new era in astronomical discovery. News of the Dutch invention had spread through Europe like wildfire in the early part of 1609. Telescopes constructed from badly made spectacle lenses were being offered for sale at fabulous prices, even though the views through them were blurred. At the University of Padua, where he was professor, Galileo had set up a workshop for making scientific instruments and had acquired a deserved reputation for skilled craftsmanship. In the space of a few weeks, Galileo carefully ground lenses from the finest Venetian glass and built the best telescope in the world. It brought him instant international fame and was the first of many to be manufactured by his workshop.

Galileo began a survey of the heavens in 1609 using a telescope that gave him a magnification of 30 times. He was the first person to direct a telescope skywards and make a record of what he saw. Wherever he turned his gaze, new and amazing sights greeted him. A family of moons belonging to the planet Jupiter was one of the most significant. With a series of observations made between January and March 1610, Galileo demonstrated that four bright 'stars' near to Jupiter were not stars at all but moons orbiting around the planet. This discovery was not merely of scientific interest. It was political dynamite! It was a powerful piece of evidence in favour of a Sun-centred planetary system, which contradicted the religious dogma of the time. Nicolas Copernicus's heliocentric theory had been in circulation since 1543 but had not been generally accepted. Its lack of appeal was partly because it undermined the authority of the church and partly because it did not square with actual observations of the planets and the philosophical reasoning prevailing at the time. One of the arguments against Copernicus claimed that the Moon would be left behind if Earth moved. Newton's theory of gravity would not be published until 1684, so the concept of an attractive force to keep moons tied to their planets was some years off. Now here was Jupiter, indisputably going around the Sun, with moons that did not get left behind.

Galileo recorded his first observations of Saturn in July 1610. Having found four moons in orbit around Jupiter, he must surely have been on the lookout for satellites of Saturn. But either he failed to detect Titan, or he did not recognise it as a moon of Saturn. His telescopes may not have been good enough to discern the dim reddish

Figure 1.2. Three sketches of Saturn made by Galileo in 1612. He thought that the appendages he could see on either side of the planet might be stationary moons of some kind and never realised that Saturn was surrounded by a ring system.

speck, or he may have been led into a blind alley by the puzzling appearance of the ringed Saturn. Galileo thought that the appendages he detected on either side of Saturn were moons of some kind, though they were clearly different in character from Jupiter's moons. Galileo remained baffled throughout his life. He never realised that his apparently triple Saturn was in reality a planet surrounded by a set of rings. The privilege of discovering Saturn's largest moon was to fall to a gifted young Dutchman who would ultimately earn a reputation as one of the greatest scientists of the seventeenth century. His find was no quirk of chance but the reward for a major advance in telescope making.

Luna Saturni

Christiaan Huygens discovered the moon we now know as Titan on the 25th of March 1655. He announced the find publicly a year later, in a pamphlet called *De Saturni luna observatio nova*. The telescope he used was cumbersome by modern standards but in 1655 it was a technical breakthrough. In collaboration with his brother Constantyn, Christiaan Huygens developed a machine for grinding and polishing lenses that made use of gears. Until the Huygens' invention, lenses were all ground and polished by hand. The process was laborious and it had proved very difficult to make the gently curved long-focus lenses that gave the least distortion in a telescope. Using their new machine, the Huygens brothers experienced little difficulty in producing long-focus lenses. The first to be incorporated as the main lens of a telescope had a focal length of about 3.6 m. This meant the telescope had to be almost 4 m long. A closed tube was out of the question and the main lens was mounted high up on a pole (Figure 1.4). A lens giving a magnification of 50 times served as an eyepiece. When Huygens turned his lanky telescope on Saturn, he noted a small point of light close enough to the planet to raise suspicions of an association

Figure 1.3. Christiaan Huygens (1629–1695), who discovered Titan in 1655.

between the two. Observing on subsequent nights, he saw the speck of light complete a circuit around Saturn over a period of 16 days. Innovation had paid off.

Englishman Christopher Wren, better known as an architect than for his early research in astronomy, and the German Johannes Hevelius both testified later that they had observed Titan through their telescopes before Huygens did but had never suspected it was anything other than a background star. The history of astronomy is full of such sorry tales of 'pre-discovery' observations by individuals who have lived to regret their lack of perception, or died without ever knowing what they missed. Indeed, Galileo never knew he had seen the planet Neptune.

Huygens had a rare combination of talents. Like Galileo, he was a practical inventor and skilled craftsman as well as being a brilliant mathematician and prolific writer. One of his greatest achievements was developing the wave theory of light but, in the late 1650s, between observations of Saturn, he was busy inventing and perfecting the pendulum clock. All the same, he found time to write *Systema Saturnium*, which was published in 1659. Saturn's mysterious appendages had remained unexplained since Galileo first reported them in 1610.

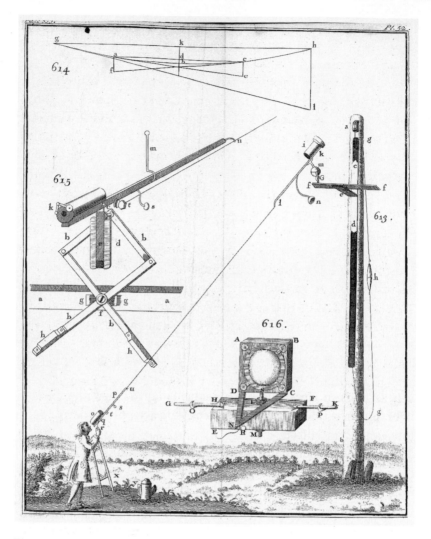

Figure 1.4. The 'aerial' telescope that Huygens was using when he discovered Titan, as illustrated in a treatise of 1738, *Compleat System of Optics* by R. Smith. Photograph courtesy of the Royal Astronomical Society.

Struggling with telescopes that were not up to the job, the handful of observers who tried simply could not make sense of what they were seeing. With the advantage of the superior telescopes he made for himself, and a brilliant mind, Huygens resolved the mystery once and for all. A thin flat ring around Saturn's equator could explain every feature of the telescopic observations recorded in the previous 39 years. The discovery of Titan orbiting Saturn in line with the 'appendages', Huygens said, was the key that led him to his correct conclusion. It was to be 198 years before another great physicist, James Clerk Maxwell, proved that the rings must themselves consist of countless miniature moonlets.

To Huygens, the world he discovered was both faceless and nameless. His seventeenth century telescopes were not remotely capable of seeing Titan as a disk and the idea of giving moons individual names didn't catch on until the middle of the nineteenth century. Huygens referred to the moving point of light simply as Luna Saturni – Latin for 'Saturn's moon'.

Over the next 30 years, the tally of moons was raised to Jupiter 4, Saturn 5. But for a long time that was it. No more moons were discovered anywhere in the solar system until 1787 and the question of names never arose. For over 100 years, the four moons of Jupiter and the five moons of Saturn were designated by Roman numerals, in order of distance from their parent planet. The system was practical enough, if unimaginative. Then in 1787, William Herschel spotted the satellites of Uranus we now call Oberon and Titania. Two years later he followed them up with two more moons of Saturn. Inconveniently, they were nearer to Saturn then the other five. What was to be done? Numbering them VI and VII would play havoc with the ordering system but re-numbering all the moons would lead to worse confusion. Not surprisingly, confusion reigned.

William Herschel's son John came up with the solution. Give moons names. Then whatever numbering system is finally adopted, at least each world is individually identified. The discovery in 1848 of Saturn's eighth moon, also out of keeping with the original order, clinched the matter. The astronomical world gratefully accepted John Herschel's names. Drawing on Greek mythology for names connected with the god Saturn – or Cronus as his equivalent deity was known in Greek – Herschel gave us Mimas, Enceladus, Tethys, Dione, Rhea, Titan, Hyperion and Iapetus. In some ways, Titan was a strange choice. Unlike the others, it was not the name of an individual but the collective name for six of the male offspring of Uranus and Gaia. Cronus, Hyperion and Iapetus were Titans. Mimas and Enceladus were two of 24 giants who were also brothers to Cronus. The female deities Tethys, Dione and Rhea were his sisters.

Dark ages

Titan's existence was known and its period of revolution around Saturn roughly determined but that was virtually the state of knowledge about Titan for more than 200 years. While increasingly powerful telescopes opened up unimagined vistas on the universe at large, the moons circling the planets of the solar system – Titan included – remained as diminutive points of light, stubbornly beyond reach. Astronomers largely turned their attention to other things, while research on moons was forced to endure long dark ages. Enlightenment would not really arrive until the space era.

Even building the list of saturnian moons proved to be a tedious process. Huygens discontinued his search after finding Titan in 1655. A quaint belief in numerology apparently led him to conclude that the inventory of the solar system must be complete. Unscientific reasoning of that kind by someone of such ability seems extraordinary from our perspective in the twenty-first century but this was an era when people ardently looked for divine harmony in the construction of the heavens. In the event, Huygens' subsequent lack of interest left the way clear for Giovanni Domenico Cassini.

Cassini came from Italy but in 1669 he was lured to Paris by King Louis XIV to direct the first observatory there. He became a French citizen in 1673 and was known afterwards as Jean-Dominique Cassini. With a relatively modest telescope he discovered Iapetus in 1671 and

Figure 1.5. Giovanni Domenico Cassini (1625–1712). He discovered the moons of Saturn now known as Iapetus, Rhea, Dione and Tethys, and the Cassini division in the ring system. After becoming a French citizen in 1673 he was known as Jean-Dominique Cassini.

Rhea in 1672. Using a more powerful instrument, he brought the total of known saturnian moons to five with his 1684 discovery of Dione and Tethys. Cassini was also the first to draw attention to the dark gap in Saturn's rings now universally known as the Cassini Division.

Next to add to the slowly growing catalogue of Saturn's satellites was William Herschel. Few celestial targets escaped the attention of this eagle-eyed musician-turned-astronomer from Hanover, who became internationally famous after discovering Uranus in 1781. In 1789, Saturn's rings appeared edge-on as viewed from Earth. In effect, they virtually disappear from view for several weeks. This made it easier for Herschel to detect for the first time two faint inner satellites that would be known as Mimas and Enceladus. But that was not all. Herschel made numerous observations of all seven of the moons he was aware of, computing the time each one took to make a revolution around Saturn. With further refinements from other observers, Titan's period of revolution was pretty well determined by the 1840s. The German astronomer Friedrich Wilhelm Bessell quoted 15 days 22 hours 41 minutes 24.86 seconds, only about half a second out from the modern accepted time.

In the nineteenth century, the study of how heavenly bodies move – the science of celestial mechanics – became highly sophisticated. Minute variations in the courses of the planets and their moons could be explained when the small gravitational tugs they each give the others were taken into account. John Couch Adams in England and Urbain J. J. Leverrier in France predicted that there was a planet beyond Uranus by assuming that its gravitational influence was responsible for pulling the errant Uranus off course. Not only that, they pinpointed where the unknown planet would be found. The discovery of Neptune in 1846, just as predicted, was a triumph for the mathematicians. In similar fashion, there were data to be gleaned about Titan's mass from the way it pulled its fellow moons around.

Saturn's eighth moon, Hyperion, proved to be the vital key in this ingenious exercise. Hyperion was discovered independently by William C. Bond at Harvard University and the noted English amateur, William Lassell, in September 1848. By an amazing coincidence, both became satisfied that they had found a new saturnian satellite on the night of the 19th. Hyperion turned out to be Titan's nearest neighbour in the saturnian family and it soon materialised that the pair interact in a remarkable way. To explain how, we should take a closer look at their orbits.

The orbits of all planets and moons are elliptical but some are a great deal more elliptical than others. Although Titan's orbit is close to being circular, Hyperion's is noticeably elongated. Saturn sits off centre, so Hyperion's distance constantly changes, swinging between 1.33 million km at one end of its orbit and 1.64 million km at the other. Titan meanwhile ranges between 1.11 and 1.26 million km from Saturn. The two orbits are not tilted to each other. Titan's is nested inside Hyperion's, like rings on a target.

Now imagine lining up Titan and Hyperion on the same side of Saturn. We put them where Hyperion is at its greatest distance from Saturn, so they are both on the long axis of Hyperion's orbit. The starting gun fires and the two take off like athletes racing around a track, except that this is an unfair competition. According to the laws of planetary motion, the more distant a moon lies from its planet the slower it's forced to go. Titan will lap Hyperion sooner or later. In fact, Titan catches Hyperion when it's made exactly four circuits, about 64 days later. What's particularly interesting is the fact that the pair are virtually back at the starting blocks, because Hyperion has done three laps in the same interval of time. Astronomers have a technical expression for this phenomenon. Hyperion and Titan are said to be in a 3:4 resonance.

Real life situations are rarely simple. In the case of Hyperion and Titan, the catch-up position isn't exactly where the two started on the long ('major') axis of Hyperion's orbit. But it's a welcome complexity for astronomers who would like to estimate Titan's mass. Titan's pull on Hyperion wants to bring the line-up back to the starting position. Over time, the interaction between Titan and Hyperion causes the Saturn–Titan–Hyperion line to swing from one side of Hyperion's major axis to the other, like a pendulum. The most it deviates is 36° and the time for a complete swing is 640 days. Crucially, the time-scale of this pendulum-like action in the line-up of the two moons is set by the mass of Titan. Several mathematicians worked on the motion of Hyperion in the 1880s, including the American George W. Hill. Hill determined that Titan's mass was 1/4714 that of Saturn. The 'modern' value is 1/4262, so Hill was about 20% off – not bad at all.

The incident of the Spanish eyes

There is no doubt that some individuals are gifted with exceptionally acute eyesight. Their problem is convincing everyone else – ordinary mortals with normal vision – that they really see what they claim. If no-one else is capable of seeing what you see, who is to say whether your description is genuine or not? Even if you are later proved correct, your interpretation could be put down to a lucky guess. If you seem to be claiming a feat of visual acuity way beyond most people's capability, you can expect some healthy scepticism.

In the early twentieth century, many astronomical observations were still made by eye. Observers would look through their telescopes and sketch what they saw. This is a technique still widely favoured over photography by amateur observers of the planets, and for good reason. Incessant motion within the atmosphere causes images to squirm like jellyfish. It is rather like looking at a coin on the bottom of pool stirred by the wind or swimmers. Expose the dancing image to a camera for more than a fraction of a second and you have a blur. By comparison, the trained human eye, with its fast link to the brain, is capable of capturing in memory fleeting moments of clarity. Nature has its moments of benevolence too, donating periods of relatively steady 'seeing' from time to time.

José Comas Solà observed Titan with a 38-cm telescope at the Fabra Observatory in Barcelona, Spain, on the 13th of August 1907. He wrote:

> ... with a clear image and using a magnification of 750, I observed Titan with very darkened edges (somewhat similar to those one observes on the disk of Neptune), while on the central part, much brighter, one sees two round, whiter patches, which give the appearance of a blurred

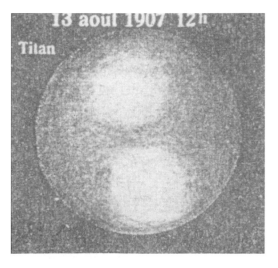

Figure 1.6. The sketch of Titan made by José Comas Solà in 1907 and published in 1908 in the *Astronomische Nachrichten* (Nr 4290, p. 287). The darkening around the periphery of the disk is believable but the two bright spots are hard to explain.

double star. We may suppose reasonably, that the darkening of the edges demonstrates the existence of a strongly absorbing atmosphere around Titan.

It is the first hint that Titan has an atmosphere. But could he have genuinely made such an observation? The claim certainly stretches credibility to its limits. At the time it was not generally taken seriously and it was never repeated. But it was there in print. Thirty-six years later, indisputable evidence of an atmosphere around Titan would raise the intriguing question of whether Comas Solà was the first to discover it. By then the Spaniard had been dead for seven years. With hindsight, how convincing was his evidence? Would it convince a jury in a court of law, beyond reasonable doubt?

What Comas Solà claimed to have observed was a disk fading off in brightness between the centre and the edge, making it substantially darker all around its periphery. A solid planet or moon without an atmosphere, such as our own Moon, does not look like this. Its disk is bright all over, with a hard edge. We can say today that this edge darkening certainly affects Titan's appearance but it is difficult to detect even with the best of modern telescopes. Was it technically possible for Comas Solà to see it by eye with his modest telescope?

We have to bear in mind the apparent size of Titan's disk. Titan is 5150 km across, say 5400 km when we add the substantial thickness of its hazy atmosphere. But Saturn is typically ten times farther from the Sun than Earth. When Comas Solà observed Titan in August 1907 it was more than 1.3 billion km away. Its disk is so small that 2000 Titans in a row would just stretch from one side of the Moon to the other. It was rather like looking at a golf ball 15 km away.

Even with perfect conditions in the atmosphere (or no atmosphere at all), the level of detail someone with normally good eyesight can distinguish is still limited – by the physical dimensions of the telescope. A telescope with a large main collecting mirror or lens can resolve finer details than a smaller telescope. A simple mathematical formula gives the size of the smallest features a particular telescope can separate. The calculation for Comas Solà's telescope says the smallest things it can distinguish are half or three-quarters the apparent size of Titan. In other words, Comas Solà was operating at the absolute limit of what his telescope could do when he spotted lighter and darker areas on Titan. Even under the best of conditions he shouldn't have been able to see much.

However, there is some persuasive evidence that Comas Solà's visual powers did verge on the superhuman. In the report that contains the Titan observations, he presents drawings of Jupiter's satellites Callisto and Ganymede. The broad features he sketched compare reasonably well with what we now know about the surfaces of these moons. He also suggested that Jupiter's closest large satellite, Io, looked flattened, pulled, he supposed, into a lemon shape by Jupiter's gravity. Although Jupiter's gravity has overwhelming consequences for Io, driving violent volcanism, Io isn't deformed nearly enough to see by eye. However, Io's polar regions are darker than the rest of its surface, so it may well have looked distorted to Comas Solà. Does this prove that all his observations were genuine? No-one can be certain what the verdict should be but the evidence for the defence cannot be ignored. In science, though, being right often isn't enough: you have to be right in such a way that other scientists can check that you're right. Since no-one else could verify the observation it is difficult to argue that Comas Solà 'discovered' the atmosphere.

Nevertheless, at least one person whose opinion counted regarded the report as credible. He was the distinguished British astrophysicist Sir James Jeans. Jeans's first book, *Dynamical Theory of Gases*, was initially published in 1904, before Comas Solà's observation but, in an edition published in 1925, Jeans applied the theory to Titan.

The dynamical (or kinetic) theory of gases says that many aspects of the behaviour of a gas are predictable if one imagines it as a collection of molecules whizzing about at speed, constantly colliding and bouncing off each other in a manner resembling a fantastic game of three-dimensional snooker. The hotter the gas, the faster the molecules go.

Without something to confine it – a box for example – a gas will quickly disperse as the molecules shoot off in every direction. In an atmosphere around a planetary body, the gas molecules are pulled towards the planet by gravity. Gravity acts rather like a box, but a leaky one without a lid. If any of the molecules near the top of the atmosphere gather enough speed to exceed the escape velocity, they can disappear into space and be lost from the atmosphere for ever. So there's a balancing act between the gas's natural tendency to spread out in all directions and the restraining force of gravity. A body has the best chance of hanging on to its atmosphere if (a) it's massive – so its gravity is strong, (b) the gas molecules are themselves relatively heavy, and (c) it's cold – so the molecules generally move slowly and don't reach escape velocity.

Jeans noted that, in the frigid depths of the solar system where Saturn and its moons orbit the Sun, Titan's gravity would be strong enough to retain an atmosphere for as long as the solar system has existed. His decision to introduce the case of Titan was no coincidence. Here was a splendid opportunity to put the kinetic theory of gases to work. Though Jeans says that an atmosphere has been discovered, frustratingly, he does not give his source of information, but it's fair to presume it must have been Comas Solà. Jeans's calculations showed that light gases, such as hydrogen and helium, would easily escape. If Titan was proved to have an atmosphere, it was going to consist of heavier molecules. Leading candidates included argon, neon, nitrogen and methane.

The methane fingerprint

With rare exceptions, astronomers can never hope get their hands on material from the objects they study. But even though samples of matter are out of the question, energy samples arrive unsolicited. Light and its invisible relatives – ultraviolet and infrared radiation and radio waves, for example – are there for the taking with the right telescopes and instruments. Fortunately, many of the substances in stars, planets, or indeed anything that shines, indelibly impress their unique fingerprints on the energy they radiate away.

Extracting the tell-tale fingerprints from a beam of light is achieved by spectroscopy – a process rather like disentangling a vast skein of multicoloured yarn of various lengths. Each colour corresponds to a different wavelength and the length of thread represents the intensity of the light in that individual colour. The signature of a particular chemical is a unique combination of threads with certain lengths and colours.

For scientific analysis, a spectrum has to be transformed into a trace, displaying the ups and downs of light intensity with wavelength. Common features in such spectrum traces have their own descriptive expressions in the scientific jargon. A series of closely spaced narrow dips, which may appear to be merged together to make a single broad dip, is an 'absorption band', for example. The fingerprints of molecules typically take the guise of absorption bands.

In the winter of 1943–44, the Dutch–American astronomer Gerard P. Kuiper used the 82-inch (2.08-m) telescope at the McDonald Observatory in Texas to record the spectra of the ten largest moons in

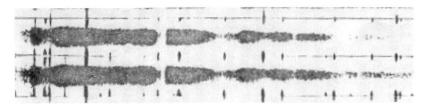

Figure 1.7. Two of Gerard Kuiper's photographic spectra of Titan, taken in the winter of 1943–44 with the 82-inch (2.08-m) telescope at the McDonald Observatory in Texas. They were published in the *Astrophysical Journal* in 1944 (vol. **100**, p. 378). Wavelength increases from left to right. They are negatives, so the methane absorption bands, which disclosed the presence of an atmosphere on Titan, appear as light gaps in the broad, dark, horizontal strips. The short vertical lines are wavelength reference marks.

the solar system in both visible light and the near infrared. This exercise was challenging at that time and had not been attempted before. The telescope was still relatively new, having been completed in 1938.

Kuiper's spectrum of Titan immediately stood out from the rest. Uniquely, it contained absorption bands identified with methane gas. Titan's orange hue was also apparent in the data, confirming what many an observer had noted by eye. Though a link with the atmosphere was an obvious connection to make, Kuiper could not know at that time whether the surface of Titan or the atmosphere itself was responsible for Titan's distinctive colour. Kuiper published his results in a paper under the title *Titan: A satellite with an atmosphere*. While James Jeans had shown decades earlier that physics and chemistry allowed Titan to have an atmosphere, the proof of its existence was still for many a startling revelation.

Unlike Comas Solà's obscure and unrepeated observation, Kuiper's evidence was clear and indisputable: Titan was no run-of-the-mill satellite. Yet the strong signature of methane was far from the last word on Titan's atmosphere. The bigger story would have to wait until 1980 and the arrival the *Voyager 1* spacecraft. Now we know that methane, which trumpets its presence with a strong spectral signature, accounts for a mere few per cent of Titan's atmosphere at most. Its publicity-shy but more abundant partner turned out to be nitrogen. As in Earth's atmosphere, the nitrogen atoms pair up to form molecules, N_2. Together, nitrogen and methane weigh down on Titan with a pressure one and half times greater than atmospheric pressure on Earth.

Voyager 1 reached Saturn in November 1980 after a journey lasting just over 13 years. Its encounter with Titan was brief but intimate.

The craft closed in to a mere 4394 km from the surface at nearest approach, its camera and full arsenal of instruments trained on the mysterious moon. In the 1970s, before *Voyager 1*'s arrival, astronomers began to suspect that Titan's atmosphere contains clouds and haze. Yet they held out the hope of gaps – a glimpse perhaps of the surface below. But it was not to be. *Voyager*'s camera, sensitive to visible light, returned images of a moon comprehensively swathed in a global blanket of orange smog. It was a disappointment but not so great a surprise. Ultraviolet light from the Sun breaks up the molecules of methane and nitrogen in Titan's upper atmosphere, releasing the ingredients to cook up a soup of complex chemicals. Some of the new substances created from the dismembered fragments are probably polymers – large, chain-like molecules. According to *Voyager 1* data, the dark particles suspended high above Titan's surface are about 0.2 to 1.0 micron across. No-one knows the details of their chemistry for sure. Whatever their nature, they are guilty of concealment and of provoking intrigue on another world a billion miles away.

A singular satellite

Titan's atmosphere is distinctive, fascinating and unique in the solar system, imbuing its owner with the qualities of a true planetary world according to all our preconceptions of what moons and planets should be like. Reinforcing this notion of planetary status, the composition of Titan's nitrogen-rich atmosphere is beguilingly similar to Earth's and unlike every other substantial atmosphere in the solar system. Not only that, the pressure and density of the atmospheres surrounding Earth and Titan are similar. No other world's atmosphere matches Earth's so closely in this respect. But there the similarities between Earth and Titan end. The trace gases in the two atmospheres are very different, largely because of the low temperature on Titan.

As a moon of Saturn, Titan is firmly in the 'outer' solar system. This means that it is cold – very cold. Its surface is 180 °C below zero – or 94 K on the absolute temperature scale preferred by scientists. It means also that Titan is made of a mix of ingredients different from those familiar on Earth and the other terrestrial planets, which are made almost entirely of rock and metal. Titan, like most of the other bodies in the outer solar system, is composed mainly of ice. Carbon dioxide and water vapour, minor but significant gases in Earth's atmosphere, are frozen solid on Titan. Instead, Titan's atmosphere boasts a cocktail of carbon-based chemicals, including ethane, acetylene and carbon monoxide.

In some respects Titan is like the planet Venus. Both have atmospheres that are thick and opaque to sunlight and both rotate slowly. This combination appears to lead to fast winds in the upper atmosphere, which zip around from east to west. In another sense, Titan is faintly reminiscent of Mars, in that the tilt of its equator results in pronounced seasons and the movement of atmospheric gas from the summer hemisphere to the winter one. On Mars, where the atmosphere is thin, the effect is huge: carbon dioxide frost evaporates in the summer hemisphere and snows out on the winter hemisphere. Atmospheric pressure changes by around 30%. On Titan, with its much thicker atmosphere, the effect is much more subtle. The haze high in Titan's atmosphere seems to be driven from the summer hemisphere to the winter one – changing Titan's brightness quite dramatically as it moves.

The magnificent seven

Larger than Mercury and Pluto, Titan was thought for a long time to be the most sizeable satellite in the solar system but *Voyager 1* set the record straight. As it turned out, Titan's solid globe, with a radius of

Figure 1.8. A montage of the seven largest planetary satellites in the solar system (see Table 1.1). From left to right, top row: the Galilean moons of Jupiter, Ganymede, Callisto, Io, Europa; bottom row: the Moon, Titan, Triton. NASA images. (In colour as Plate 2.)

Table 1.1. *The 'magnificent seven', Mercury and Pluto*

Name	Moon of	radius (km)
Ganymede	Jupiter	2634
Titan	Saturn	2575
Callisto	Jupiter	2403
Io	Jupiter	1821
The Moon	Earth	1738
Europa	Jupiter	1565
Triton	Neptune	1353
Mercury	—	2439
Pluto	—	1150

2575 km, is fractionally smaller than Jupiter's moon Ganymede. Titan appears superficially larger than Ganymede because of its thick atmosphere rendered opaque with haze.

A total of seven satellites in the solar system surpass Pluto in size. Ganymede and Titan head the list of this 'magnificent seven'. The remaining Galilean moons of Jupiter – Io, Europa and Callisto – account for a further three. The other two in the club are our own Moon and Neptune's satellite, Triton. Size sets these seven in a class on their own above the other 60-odd (a number that increases year on year). Their nearest challengers come in with radii under 800 km.

The four large planets of the outer solar system each play host to a substantial swarm of satellites. Many of these moons are small irregularly shaped chunks, their dimensions reckoned in tens of kilometres at the most. A proportion, we can be sure, are captured asteroids, trapped by gravity after some chance close encounter aeons ago. But others, Titan included, most likely came into being where we observed them today, condensing out of a disk-like nebula surrounding their nascent parent planet. Each of these systems of satellites is like a solar system in microcosm.

In the very early solar system, as now, temperature declined with increasing distance from the Sun. The local temperature had a profound effect on the final composition and structure of the different planets. Nearest the Sun, scorching Mercury is dense rock and metal, and bone dry. Farther out, Earth has more rock and less metal and its surface is awash with vast oceans of water. In the much cooler environment of the outer solar system, the giant planets accumulated great atmospheres of light, volatile gas.

When Jupiter and Saturn first condensed out of the solar nebula, they were hot. As sources of heat, these bodies had an effect on their developing satellite systems similar to the Sun's on its emerging planets. That effect is very obviously reflected in the bulk composition of the four Galilean moons of Jupiter.

Closest to Jupiter, Io seems largely bereft of water. From what we can see, Io is made of rock and sulphur, although its density implies that it has a core of iron. Next out is Europa. Although this is the brightest of the Galileans, its icy crust is a thin veneer: its density requires that most of its interior is rock and that only the outer 170 km or so is water in solid or liquid form. Because many compounds we are familiar with on Earth as volatile gases are frozen solid in the cold outer solar system, we call them ices – methane ice, ammonia ice and so on. Accordingly, planetary scientists often specify 'water ice' when they mean frozen water, since the simple word 'ice' is ambiguous. The surface of Europa is water ice but there is a lot of evidence to suggest that, a few kilometres down, the water is liquid. There seems to be a global subsurface ocean on which the ice crust floats and grinds. Certainly there are places where it looks as if water has welled up through the ice and areas where rafts of broken ice have moved around. We can see how, in principle, many of the pieces could be reassembled like a jigsaw puzzle. Tidal energy supplies the heat to power Io's volcanism and melt Europa's ice.

By contrast with Europa, both Ganymede and Callisto have more ice. The proportions of rock and ice are reflected by average density: more ice and less rock and metal means a lower density.

Unexpectedly, magnetic measurements made by the *Galileo* spacecraft suggest that Callisto also has a subsurface water layer deep below its icy crust. Ganymede probably has one too, although its signature is hidden by Ganymede's own magnetic field. This came as something of a surprise; while Europa is close to Jupiter and is kneaded by tidal forces, Ganymede and Callisto are much further out, so should experience much less tidal heating. If Callisto has an internal ocean, then it raises the possibility that Titan might have one too. Titan also has the likely advantage of abundant ammonia acting as an antifreeze in its interior.

Table 1.2. *Density of Jupiter's Galilean moons*

Moon	Distance from Jupiter (km)	Density (g/cm³)
Io	421 600	3.53
Europa	670 900	2.99
Ganymede	1 070 000	1.94
Callisto	1 883 000	1.85

Meet the family

While Jupiter has four comparably large moons, Titan dwarfs all its siblings. The effects of Saturn's early life as a warm planet are less obvious among this tribe than in the case of Jupiter. With only one third Jupiter's mass, the coalescing of Saturn generated substantially less heat and the whole saturnian system is in any case twice as far from the Sun. As a result, all Saturn's moons are icy without distinction by distance from their parent but we can expect that there were subtle effects on their composition. Certainly, the overall cooler regime around Saturn can explain why Titan successfully formed and retained an atmosphere, when the Galilean moons of Jupiter did not. So, what of Titan's fellow moons? What details we know of them stem almost entirely from the encounters of *Voyagers* 1 and 2. At the time we write, names have been allotted to 18 and 12 discovered in 2000 have provisional designations. Studies of the way the rings are organised strongly suggest the presence of another but it has so far not been sighted.

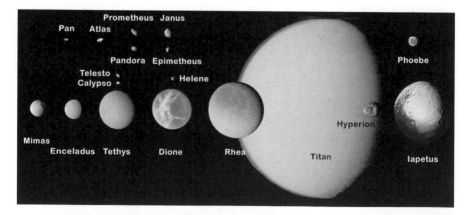

Figure 1.9. A montage of *Voyager* images of saturnian moons. NASA images. Adapted from a montage prepared by David Seal. (In colour as Plate 3.)

Stepping outwards from Titan, we come first to the irregular rocky world of Hyperion, 300–400 km across. At that size, it could just have pulled itself into a sphere, had it been assembled like most planets and moons. Its potato-shape demands an explanation. One possibility is that it's a chunk left over from a catastrophic impact, the largest remnant of something that was once considerably bigger. Most satellites, as is the case with our own Moon, always present the same face towards their parent planet, locked in a permanent relationship as a result of tidal interactions over the ages. Not so Hyperion. This satellite's oddball shape forces it to tumble such that its orientation cannot be reliably predicted, a case of mathematical chaos.

Next out is Iapetus. This world is a third the size of Titan and has two contrasting faces – one bright and icy, the other dark. Iapetus always presents the same face to Saturn. This means that the side pointing forwards along its orbit is always the same and, of course, the opposite side faces backwards. The forward pointing, or leading, face is the side covered by the mysterious dark material. As viewed from Earth, when Iapetus is on one side of Saturn we see its bright trailing face, and when it is on the other side of Saturn we are looking directly at the dark leading face. Cassini, who discovered Iapetus in 1671, soon became aware of the fact that it disappeared when on the far east side of Saturn. He correctly interpreted what was going on, writing, 'it seems that one part of his surface is not so capable of reflecting to us the light of the Sun which maketh it visible, as the other part is'.

Such a remarkable dichotomy between two sides of Iapetus has to have an explanation. The most likely is that material from Phoebe spirals in and peppers the front face of Iapetus. The dark area of Iapetus isn't quite a whole hemisphere. The dark stuff smears around the equator somewhat but doesn't appear much at high latitudes, consistent with the contamination by dust streaming from Phoebe.

On the most distant outskirts of the saturnian system, we come to the swarm of small satellites discovered in 2000 and Phoebe. Phoebe is more than ten times farther from Saturn than Titan. Apart from their orbits, little is known about any of these tiny worlds, except that Phoebe is irregularly shaped and dull red. All have highly inclined orbits. Several, including Phoebe, follow retrograde orbits – that is, they travel backwards in relation to the direction taken by the majority of moons. It is hard to imagine how bodies that formed in a disk of dust and gas with the rest of the satellites could have ended up in such

Table 1.3. *The satellites of Saturn*

Name	Discoverer	Year of discovery	Mass (kg)	Radius (km)	Mean density (kg/m³)	Semi-major axis of orbit (km)	Sidereal orbit period (days)	Orbital inclination (°)	Orbital eccentricity
Pan	M. Showalter/ *Voyager 2*	1990		10		133583	0.575		0.000
Atlas	R. Terrile/*Voyager 1*	1980		19×17×14		137670	0.602	0.3	0.003
Prometheus	S. Collins/*Voyager 1*	1980	1.4×10^{17}	74×50×34	270	139353	0.613	0.0	0.002
Pandora	S. Collins/*Voyager 1*	1980	1.3×10^{17}	55×44×31	420	141700	0.629	0.0	0.004
Epimetheus	J. Fountain *et al.*/ *Voyager 1*	1980	5.4×10^{17}	69×55×55	630	151422	0.694	0.34	0.009
Janus	A. Dollfus	1966	1.92×10^{18}	99×96×76	650	151472	0.695	0.14	0.007
Mimas	W. Herschel	1789	3.75×10^{19}	199	1140	185520	0.942	1.57	0.020
Enceladus	W. Herschel	1789	6.5×10^{19}	249	1000	238020	1.370	0.00	0.005
Tethys	G. Cassini	1684	6.22×10^{20}	530	1000	294660	1.888	1.86	0.000
Telesto	B. Smith *et al.*/ *Voyager 1*	1980		15×13×8		294660	1.888	~0	~0
Calypso	D. Pascu *et al.*	1980		15×8×8		294660	1.888	~0	~0
Dione	G. Cassini	1684	1.1×10^{21}	560	1500	377400	2.737	0.02	0.002
Helene	P. Laques & J. Lecacheux	1980		18×16×15		377400	2.737	0.0	0.005
Rhea	G. Cassini	1672	2.31×10^{21}	764	1240	527040	4.518	0.35	0.001

Titan	C. Huygens	1655	1.345×10^{23}	2,575	1881	1221830	15.945	0.33	0.029
Hyperion	W. Bond & W. Lassell	1848	1.5×10^{19}	$185\times140\times113$		1464100	21.277	0.43	0.104
Iapetus	G. Cassini	1671	1.59×10^{21}	718	1020	3560800	79.330	14.72	0.028
S/2000 S5	B. Gladman	2000				11365000	449.22	46.2	0.334
S/2000 S6	J. Kavelaars & B. Gladman	2000				11440000	451.48	46.7	0.322
Phoebe	W. Pickering	1898	7×10^{18}	110		12944300	544.23	175.3	0.163
S/2000 S2	B. Gladman	2000				15199000	686.92	45.1	0.363
S/2000 S8	J. Kavelaars & B. Gladman	2000				15645000	728.93	152.9	0.270
S/2000 S11	M. Holman & T. B. Spahr	2000				16392000	783.30	34.0	0.479
S/2000 S10	J. Kavelaars & B. Gladman	2000				17611000	871.17	34.5	0.473
S/2000 S3	B. Gladman & J. Kavelaars	2000				18160000	893.07	45.6	0.295
S/2000 S4	J. Kavelaars & B. Gladman	2000				18239000	925.70	33.5	0.536
S/2000 S9	B. Gladman & J. Kavelaars	2000				18709000	951.38	167.5	0.208
S/2000 S12	B. Gladman & J. Kavelaars	2000				19470000	1016.83	175.8	0.114
S/2000 S7	B. Gladman & J. Kavelaars	2000				20470000	1088.89	175.8	0.466
S/2000 S1	B. Gladman	2000				23096000	1312.37	173.1	0.333

orbits. The most likely hypothesis is that Phoebe and its companions are late additions to the family of saturnian satellites – adopted strays.

Moving inwards from Titan, we first encounter Rhea, the largest of Titan's fellow moons though only one third as big. Rhea is literally saturated with craters. It seems there was more than one episode when impacts bombarded the saturnian system – not just the trickle of comets and asteroids from the rest of the solar system but also something peculiar to Saturn. One persistent hypothesis suggests that whatever body it was that broke up and created Hyperion in the process, shed its debris all over the saturnian system. This idea has an interesting implication for Titan. As a large satellite next door to Hyperion, it could have acquired prodigious amounts of this material. The debris could be kilometres thick and it would have saturated Titan's surface with impact craters.

Four more moons of intermediate size reside within Rhea's orbit: Dione, Tethys, Enceladus and Mimas. These predominantly icy globes are all cratered to a greater or lesser extent. Mimas is most heavily cratered, one particularly huge crater dominating the landscape of its leading hemisphere. With this striking feature, Mimas resembles the 'Death Star' from the movie *Star Wars*. Dione, Tethys and Enceladus all display evidence of watery eruptions that at some stage in their history wiped out craters and variously created plains, grooves and canyons. Dione shares its orbit with tiny Helene, which orbits 60° ahead of it. Tethys has two diminutive orbital companions, Telesto and Calypso. Unlikely as it seems at first sight, these shared living arrangements are stable. The orbits of Enceladus and Mimas actually lie inside Saturn's ring system, Enceladus close to the centre of the broad but dim E ring. A further six – possibly seven – small moons, all within the ring system, complete the family round-up.

The scars on this assorted crew bear witness to the upheavals and violence that rocked the Saturn system long ago. Reading their inscrutable faces offers us a few clues as to what occurred and what might be found beneath Titan's veil. But most likely, when Titan's secrets are finally out, there will be plenty of surprises.

Fact and fantasy

Any place about which little is known, becomes the abode of monsters. 'Here be Dragons' wrote the drafters of medieval maps of the world. With Earth pretty well explored, twentieth-century adventure writers looked farther afield to find plausible settings for the creatures born of their fertile minds. Every real planet and moon, as well as countless imaginary ones, have had their quota of aliens. In science fiction's early days, Venus, our shrouded sister world and neighbour in the solar system, was a convenient hideout. Ignorance of what it is really like was a blessing. Then, in the 1970s it became clear that Venus was too hot and hostile a planet to support any kind of life. The horizon of ignorance and menace receded farther from the Sun.

Titan, known since 1944 to have an atmosphere, made a convenient 'mystery planet' from which aliens might plausibly come. In Robert A. Heinlein's 1951 offering, *The Puppet Masters* , for example, evil slugs from Titan exert mind control over hapless humans. But Titan itself doesn't come into the story.

It was only in the 1970s, as more was learned about Titan, that it became a real backdrop for science fiction. Two novels are prominent. Arthur C. Clarke's *Imperial Earth* , written in 1976, featured the latest ideas about Titan's surface. Clarke got much of his contemporary information from Carl Sagan. Notably, the surface temperature was just a few tens of degrees below freezing, supposedly raised by a strong greenhouse effect. Among the exciting ideas are the waxworms – a sort of volcanic event – and the methane monsoon. Clarke features a large hydrocarbon lake in his story, named Loch Hellbrew. Perhaps such a feature will actually be discovered. He describes the noise of the wind on Titan, augmented by the howl of spacecraft scooping up gas in Titan's upper atmosphere for use as propellant.

Clarke is prescient in selecting Titan as a centre of power in a solar-system-wide empire – its huge reservoir of volatile gases, relatively accessible around a low-gravity body will make it an attractive fuelling stop, if not an exciting destination in its own right. However, as far as the story is concerned, Titan is simply a picturesque backdrop and the action soon moves to Earth.

A more recent novel by Stephen Baxter bears the unambiguous title *Titan* that at least points to a more pivotal role for its eponymous world. This 1997 story centres around a manned mission to Titan, a last desperate exploratory fling set against the backdrop of an increasingly technophobic and introspective society on Earth. As well as painting a terribly bleak picture of the psychological and physiological toll of long-term spaceflight, Baxter describes – in more detail even than Clarke before him – what it would be like to be on Titan. He captures many current ideas about Titan's landscape, as in the following vista from a manned spacecraft descending by parachute through Titan's atmosphere:

> And now Jitterbug drifted over a pair of giant craters, each perhaps over fifty miles across. In one of these the central peak seemed to have broadened into a dome, so that the ethane pool was contained in a thin ring around a central island. But she could see a pit at the centre of the dome, itself containing a small pool, so the whole structure had a bull's-eye shape, with the solid circle and a band of dark fluid contained by the circular crater rim. And in the second of the big craters, the outer annulus of fluid seemed to be heaped up against one wall of the crater – perhaps by some tidal effect – so that the lake was in the form of a semicircular horseshoe.

RALPH'S LOG. 1997.

I have to confess I felt a faint swell of pride when I read some of Baxter's description of Titan's landscape: the writer clearly did his homework. Compare what he has written with the summary of a paper I wrote while I was a Ph.D. student in Canterbury in 1993, published in the journal Planetary and Space Science in 1994.

> *Medium-sized craters will have central peaks, while large craters may undergo viscous relaxation, their centres doming upwards, to push the liquids into an annulus, thereby forming a ring lake. Additionally, craters with central pits may form 'bullseye lakes'. The large tidal effects of Saturn on such lakes will affect their shape, distorting rings into horseshoes . . .*

As I realised soon after writing this, and has also been pointed out to me since, tides alone will probably not cause horseshoe-shaped lakes, since both the liquid surface and the ground beneath will see the tidal forces, so there is no difference between the local horizontal on the liquid. That's certainly true for the physicist's perfect billiard-ball-smooth world. But real planets and moons have bumps and irregularities, so slopes exist for other reasons. Hence I still bet there are lots of horseshoe-shaped lakes.

Art is another medium in which proponents exploit our ignorance and indulge their speculations. The renowned space artist of the 1940s Chesley Bonnestell painted a Titan much like his impressions of the Moon, with sharp, spiky mountains. Not knowing any better, he painted a blue sky.

In the 1980s, after the *Voyager* encounter, artists impressions changed. Dull red became the dominant colour and dark sludge appeared everywhere. These aspects, while rigorously correct, do not make a particularly spectacular or appealing vista. However, in compensation, the artist can justify throwing in a few clouds and maybe a lake or two. American artist Michael Carroll made a couple of impressions, one of a robot airship cruising over the shoreline of a hydrocarbon lake, with sludge glaciers in the background, and another showing a parachute-borne probe thumping down onto an icy surface.

A perennial feature of artists' impressions of Titan is the outcome of an overwhelming temptation to show Saturn in the background. Not infrequently Saturn is shown with its rings open. The basic expression of this artistic license is perhaps arguable since at some wavelengths, Saturn should be quite visible but having the rings open is not. They would always be seen nearly edge-on from Titan.

This hard reality lays down a challenge to the artist: to imagine something aesthetically appealing that also carries the cachet of authenticity. How will artists respond when the real view from Titan's surface is finally revealed?

To Titan to know ourselves

In the 1980s, both NASA and ESA, together representing a substantial sector of the developed world, decided it was worth joining forces to go to Saturn and Titan. So why bother trekking over a billion miles, effectively costing a dollar every mile? Aside from the philosophical idea of exploring for exploration's sake, there are many sound reasons for investigating the planets, and Titan in particular – reasons that ultimately benefit everyone on Earth.

More than anything else, planetary exploration gives us a sense of perspective, a notion of who we are, where we came from and what our destiny might be. We can learn from all worlds. Each planet and moon in the solar system has its own unique history. Each is an experiment with a different set of conditions – each dealt a different hand in a geological game of chance. Tinkering with Earth is a prohibitively mammoth undertaking, which is probably just as well. We can see already the results of inadvertent human manipulation of trace gases in the atmosphere. Imagine the chaos we could cause if we actually tried! Experimenting with the bulk makeup or the interior structure of the planet would be orders of magnitude more difficult and more dangerous than tinkering with its atmosphere, although knowing what could happen is theoretically important. Yet the results of different experiments in planet construction carried out by nature are all around us, like tempting dishes (and a few burnt offerings), waiting for us to divine what the ingredients were and how they were cooked. It's not the way one might choose to learn how to cook perhaps but, by studying the results nature serves up, we at least have a hope of discovering the recipes.

Many processes that have shaped Earth only in subtle ways are stark in their brutal ubiquity elsewhere in the solar system. Impact cratering is a case in point. This important process has only been understood in the last couple of decades, largely because it does not impinge directly on human experience. In an individual human lifetime, millions of tiny meteors will flash through the sky, perhaps thousands of boulders the size of houses will burn up and explode, littering the ground with meteorites and – if the individual is unlucky – he or she might see one event like the Tunguska explosion, which flattened trees for miles in Siberia in 1908. The events that excavate great pits, like the 1-km meteor crater in Arizona, thankfully occur only every few tens of thousand years. Scarcer still are impacts producing craters 100 km or more in diameter. The formation of such a gigantic hole in the ground is accompanied by a cataclysm bringing worldwide ecological upheaval in its wake. The most recent and famous of these was the Chicxulub impact in Mexico's Yucatan peninsula, now widely believed to have caused the conditions that wiped out the dinosaurs in an episode of mass extinction.

It's difficult for humans to relate to these events. The ever-changing nature of Earth's surface obscures much of the evidence for them, so only 40 years ago, most of the craters on the Moon were thought to be made by volcanoes. Yet, after close inspection of other planets by spacecraft, we now know that virtually every body in the solar system, with the exception of the ever-seething Io, is riddled with impact craters. Only by seeing and studying the 'zoo' of different examples throughout the solar system can the details of the process and effects of large impacts be understood.

If what we believe about the evolution of the Sun and the solar system is true, then Earth's first atmosphere was lost in the convulsive blasts of wind from the nascent sun and the second is a late veneer delivered by the impacts of comets and asteroids. That atmosphere has probably remained more or less constant in total mass, although its composition has evolved in unknown ways, with an attendant variation in climate. Titan may have undergone a similar set of processes, although under different local conditions.

By studying other worlds we can get an idea of what happens when gross planetary properties change. As an example, the theory of a 'greenhouse effect' warming Earth's atmosphere was put forward as a hypothesis as early as the 1890s but the idea only really caught hold when the dramatic warming caused by the torrid pressure-cooker atmosphere of Venus was understood in the 1960s.

Instruments first developed to study Venus are now used routinely to monitor Earth's microwave emissions and to measure snow and cloud cover, rainfall and so on. No-one can say that such instruments wouldn't have been developed eventually without first being sent to the planets but space exploration spurs us on to try ambitious new ideas and less obvious approaches.

Titan is in many respects a particularly close analogue of Earth when it comes to weather. As one rises above the surface, the temperature falls, until a minimum temperature is reached at an altitude called the 'tropopause' (about 10 km up on Earth, where most airliners fly; about 40 km above the surface on Titan). The tropopause traps gases that can condense. On Earth this keeps most water vapour below 10 km, in the 'troposphere'. On Titan, methane plays the role of water: it can condense to form clouds and rain (and, perhaps, hail or snow) and may form lakes and seas on the surface. Methane is also a strong greenhouse gas. Condensable greenhouse gases introduce a powerful climatic effect, a so-called 'positive feedback' that amplifies climate changes. If the planet warms up, the atmosphere can hold more of the gas – warm air can be more humid than cold air. But if the vapour is a greenhouse gas, then having an increased supply of it can warm the surface to an even greater extent, boiling off even more vapour, and so on. For Earth at present this just makes the effects of climate change hard to predict. Only in a billion years or so, when the Sun is stronger, will the effect be so powerful as to run amok and boil off the oceans. This is something that we think happened to Venus billions of years ago. For Titan, much farther from the Sun but with a much more volatile greenhouse gas, the runaway greenhouse may be just about to happen, in geological terms.

The haze in Titan's atmosphere acts very much like ozone in Earth's atmosphere. Both are made by the action of ultraviolet light on the atmosphere (on oxygen on Earth, on methane on Titan). Both

Earth's ozone and Titan's haze block ultraviolet light from the respective lower atmospheres and surfaces – an effect whose importance has only been realised on Earth since the 1970s. Both also heat the atmosphere locally, because they absorb the ultraviolet light. There are of course big differences: on Earth ozone is part of a complicated chemical cycle – it is continuously being made and destroyed while its constituent oxygen atoms swirl in a fickle dance between oxygen, ozone, and compounds, including oxides of nitrogen and the chlorofluorocarbons released by human activity. On the other hand, Titan's haze is – as far as we know – a one-way trip for its constituent carbon and nitrogen atoms. Once it forms it eventually sinks to the surface, coating much of Titan with an unpleasant sludge.

This brief account of the climate, which we'll discuss at more length in Chapter 3, underscores why Titan is so important and interesting to study. Titan is different but not so different as to be unintelligibly alien; it is exotic, yes, but also strangely familiar. Most scientists are under no illusions about Titan being actually like the early Earth, although the comparison is all too appealing to be resisted at times. All the same, Titan will have much to teach us about the *processes* that have shaped Earth. Titan is not an Earth in deep freeze but a body that started out different in some respects (like bulk composition) yet evolved through the same processes.

Current theories of solar evolution – how the Sun's size, heat output and colour change with time – suggest that it was fainter in the past. Earth's climate should have been much colder, perhaps with ice sheets all over the planet. However, there is little geological evidence for this 'Snowball Earth'. Venus can tell us nothing on this subject: its surface was wiped clean in a paroxysm of volcanism that left nothing older than 500 million years exposed. Mars's climate history might provide some clues but it is complicated by strong seasonal and orbital variations. Mars suffers the same paradox as the early Earth – there is evidence of flowing water on the surface billions of years ago, yet it is now too cold, and the air too thin, for liquid water to flow. And if the Sun were fainter in the past, it should be even harder for conditions in the past to allow river valleys to form. Maybe Mars had a thicker atmosphere and a stronger greenhouse effect that compensated for the weak Sun; maybe Earth did too. If Titan's surface is old enough to have a record of what transpired early in the life of the solar system, then we may be able to tell whether Titan's atmosphere was thicker in the past too.

Planets are things too complex to understand in isolation – there are too many factors in play. But by studying them as a group, some patterns can emerge. Like solving a logic puzzle, or piecing together the evidence following a crime, a bit of solid information from one place can corroborate a mere suspicion from another, and slowly the pieces of the puzzle fall into place.

Tempted by fate

The fates that ordained the solar system drew a line in the sand and issued a direct challenge with Titan. They tossed us a few easy but tantalising clues: 'Here's a moon, but a special one. It's got a thick atmosphere – interesting, huh? Oh – you'll like this – there are organic compounds aplenty'. We curious humans can't resist when we are teased like that. So *Voyager 1* was targeted for a close look, answering some of the basic questions but, as ever, raising more.

The fates were at work again with another hint calculated to tantalise humans: surface conditions suggest the possibility of widespread liquids and active meteorology over a wet and perhaps exotically familiar landscape. But the surface was hidden from the *Voyagers*: 'Come back in 15 years when you've learned to see in the infrared', chuckled the fates.

Perhaps the next laugh will be ours, in 2004. The instrument packages carried by the *Cassini–Huygens* mission are so formidable that whatever wavelength, whatever way of seeing, promises the most information, it should be covered. Many of the ambiguities we are left with by remote sensing will be addressed by the *in situ* measurements made by the parachute-borne *Huygens* probe. Doubtless new mysteries will arise even as some of the old questions are answered. Somehow they always do.

Seeing Titan

Every 378 days, Saturn shines at its most brilliant in the midnight sky. Forming a line in space with Earth and the Sun, it makes a configuration astronomers call 'opposition'. At opposition, Saturn is as close to Earth as it can ever be but still almost 1.3 billion km away. Titan, we now know, is 5150 km across. This may be large as moons go but, at such an immense distance, it is shrunk to a minuscule target for an Earth-bound telescope. Deducing anything at all about this remote and secretive small world represents a significant achievement, yet astronomers have persisted until they found ways of probing through the haze to the shadowy surface below. The quest to find out from Earth what it's like on Titan's surface is an epic story. Decades of frustration made ultimate success all the sweeter.

The long and the short of it

Size matters. It's possibly the most vital of all statistics for any heavenly body. By the late 1890s, several observers had estimated Titan's size to be very roughly 2/3 of an arc second when at its nearest to Earth. It was difficult to get a more accurate figure. The problem is simply that Titan is too small to resolve. Even through a telescope theoretically large enough to distinguish details on the disk, the image is normally blurred by the shimmering of Earth's atmosphere. In fact, Titan presents us with a disk around 0.8 arc seconds in diameter but the quality of seeing from even the best mountaintop observatories rarely allows such good resolution. The American astronomer Edward E. Barnard, observing in 1894 and 1895, exploited the eye's capacity to record fleeting moments of steadiness, as José Comas Solà did a few years later. Working at the eyepiece of the 36-inch telescope at the Lick Observatory, he used a micrometer to measure directly the apparent size of Titan as best he could. Given the distance to Saturn, he inferred a diameter of 2720 miles, or about 4380 km.

In the 1940s, French astronomers Bernard F. Lyot and Henri Camichel invented a novel device for estimating the sizes of small moons and planets, aptly known as the 'disk meter'. It consisted of a small illuminated disk that could be placed in the telescope's field of view close to the actual image of the moon or planet in question. By adjusting the level of illumination, the colour of light and the size of the disk until it matched the appearance of the real object, observers could get a reasonable indication of the size and brightness of the moon or planet. Ingenious as it was, even this method was only good to an accuracy of 20% or so, however. Observing with a 24-inch telescope at the Pic du Midi Observatory in the French Pyrenees, Camichel came up with an angular size of 0.76 arc seconds for Titan, which translated into a diameter of 4960 km. Gerard Kuiper used a similar gadget on the 200-inch telescope at the Palomar Observatory in 1954. He declared the disk to be 0.67 arc seconds, corresponding to a diameter of about 4600 km. With hindsight, we can see he didn't do as well as Camichel.

In the 1970s another technique was devised. It exploited lunar occultations. During certain periods, the Moon crosses between Titan and Earth. Since the Moon has no atmosphere, its edge progressively cuts out the light from any object it passes over in a very clean and predictable way. When the object emerges again from behind the moving Moon, it is exposed bit by bit. Although Titan's disk couldn't be well-resolved during the occultation observations, the varying amount of light coming from Titan could be measured accurately. The times taken for its brightness to fall from maximum to zero at the start of the occultation, then rise again at the end, related directly to the size of its disk. Jim Elliot of the Massachusetts Institute of Technology observed lunar occultations of Titan in 1974 (published in 1975). He put Titan's diameter at 5800 km, give or take 200 km. This was significantly higher than previous estimates, suggesting that Titan's density was less than had been thought.

This time the figure had overshot the mark, for although the Moon's edge is sharp, Titan's is fuzzy and there is a problem. Is the edge where you can no longer see any light at all, or where the light is 50% as intense as at the centre of the object, or 10% ... ? How big you think it is depends on where you decide to draw the line. This dilemma would raise its head again during the analysis of Hubble Space Telescope images in the late 1990s.

Meanwhile, the small, relatively cheap *Pioneer 11* spacecraft launched in 1973 was on its way to Saturn. It arrived in 1979. The *Pioneer*s 10 and 11 were essentially scouts for the *Voyager* missions to follow. Their purpose was to help engineers develop techniques and to assess the risks of outer solar system operation. They carried relatively few instruments in a package amounting to only 30 kg.

It seems extraordinary now that *Pioneer 11* had no real camera but it did have an instrument that could be used to create an image – the Imaging Photopolarimeter (IPP). This device was basically a tube with a pair of photocells at one end. It could measure the amount of light coming from a single spot. One photocell was sensitive to blue light and the other to red. In front of them was a polarising filter that could be rotated to measure how polarised the light was. The tube was mounted so as to protrude out of the side of the spacecraft but it could be swivelled around one axis. Unlike *Voyager* and *Cassini*, its more sophisticated successors, the simple *Pioneer* maintained its stability in space by spinning. It rotated at roughly 5 r.p.m. around an axis of symmetry along which the radio beam of its main antenna was pointed, usually at Earth. The IPP, mounted nearly orthogonal to this axis, was swept around by the spin of the spacecraft like a scanner. Add to this the swivel movement and it was possible to

Figure 2.1. This image of Titan was taken by the *Pioneer 11* spacecraft on the 3rd of September 1979 from a range of 3.6 million km (2.3 million miles). It was constructed by laying together strips of data from the red and blue photopolarimeter channels. Although the signal-to-noise ratio of the individual pixels is quite good the image quality is poor, in part because of the uncertainty in assembling the mosaic of pieces. This was the first data to show one hemisphere of Titan brighter than the other. NASA image.

acquire sets of data in strips that could be laid side by side to build up pictures.

Among the scientists and engineers working on this experiment, which was led from the University of Arizona, were Tom Gehrels and Martin Tomasko. Gehrels, a Dutchman like Kuiper, was in charge of the enterprise. He would later set up the Spacewatch observatory, the first to routinely hunt for Earth-threatening asteroids. Tomasko took the lead on interpreting the scattering of light by hazes; he had recently done similar work on Venus. A young student called Peter Smith worked with him.

Smith published a short paper, succinctly entitled 'The Radius of Titan'. He had performed a similar analysis on the Galilean satellites of Jupiter, which *Pioneer* had observed during its Jupiter flyby, so he had all the necessary computer programs. His numbers were consistent with Elliot's occultation results but a little more accurate. The blue radius he found to be slightly larger than the red one, an effect of Titan's atmosphere.

All the attempts to gauge Titan's size up to this point were based on how big Titan looked, haze and all, but still no-one knew how deep the atmosphere went. Resolving that question required a technique exploiting radiation that could pass through the atmosphere – radio waves. The *Voyager 1* flyby thirteen months after *Pioneer 11* presented the essential opportunity and was set up specifically to measure Titan's atmospheric structure using the spacecraft's own radio signals as a probe. As *Voyager 1* passed behind Titan, its microwave radio signals were monitored on Earth. The subtle bending effects of Titan's atmosphere on the signals as they passed through were used to build up a profile of pressure and temperature. As the craft went behind Titan's solid surface the radio signals were abruptly blocked. The results of the radio occultation experiment put Titan's true diameter beneath its hazy atmosphere at 5150 km to within 1 km. With similar accurate measurements also registered for Jupiter's Galilean satellites, it became clear that Titan's atmosphere had misled those who believed it to be the largest moon in the solar system. That crown was yielded to Ganymede, a whole 60 km larger.

We know Titan's average density from the combination of its mass and radius. It turns out to be about 1.88 times that of water – mid-way between the densities of Ganymede and Callisto. Since Titan is denser than pure ice, it must also have a significant component of a material denser than ice – rock. In fact Titan should be between about 42% and 70% rock by mass, depending on the type of rock.

Taking the temperature

Even before Titan's diameter was pinned down, scientists were grappling with its other mysteries, including its temperature. In the late 1960s and early 1970s electronics began to revolutionise the way astronomy could be done. One of the new techniques opened up by electronics was infrared astronomy. Though infrared radiation is invisible, humans are not insensitive to it. We feel infrared as warmth on our bodies. Equally, we are emitters of infrared since everything gives out so-called thermal radiation, the nature of which is determined only by how hot or cold the emitter is. The infrared radiation from something is a prime indicator of its temperature. Microwaves and radio emission can similarly be used as celestial thermometers, particularly for things at the colder end of the scale.

If you know both the size of a planetary object and the amount of thermal radiation coming from it, you can estimate the temperature of the surface emitting the radiation. But situations are never straightforward. In the case of Titan, applying this technique with infrared radiation yielded a variety of answers, depending on which wavelength was observed. At a scientific workshop on Titan in 1973, temperatures reported ranged between 90 K and 140 K. The explanation was not hard to find. The different wavelengths were coming from different places in the atmosphere because of the way gases absorb infrared radiation. The first temperature measurements taken with radio telescopes were made in 1973 at the National Radio Astronomy Observatory in the USA by F. W. Briggs. The result was the moderately high value of 115 K, give or take some 40 K. On one hand, these measurements were less precise than the infrared ones, since these longer wavelengths are altogether more feeble and it is difficult to direct the wider beam of the radio telescope to exactly the right spot. On the other hand, looking at the radio waves was expected to probe right down to the surface, unimpeded by absorption in the atmosphere, unless it happened to be very deep.

Taken together, the results presented a real puzzle. Some parts of Titan's atmosphere seemed to be relatively warm while others were much cooler. What could this mean? One idea was that something high in the atmosphere was absorbing sunlight, in much the same way as ozone does in Earth's atmosphere. Another idea was that Titan's atmosphere is deep and, like the lower atmospheres of all the planets, subject to convection. That means that the gas circulates up and down, cooling as it rises. Under these circumstances, the lowest parts of the atmosphere should be the warmest, the exact temperature depending on the depth of the atmosphere or the surface pressure, which no-one knew at the time. This opened up the exciting possibility that Titan's lower atmosphere might be warm enough for life.

This discussion was especially pertinent in the early 1970s because space missions were ruling out Mars and Venus as hospitable abodes for life. Both schools of thought regarding Titan had solid reasoning behind them but only new information would resolve the debate. Carl Sagan summed up the situation in a short chapter in his 1974 book *Broca's Brain*: 'Which View is Correct? At the present time no-one knows'.

Supporting the hot haze idea were observations of the polarisation of light from Titan, suggesting that light was scattered several times before being reflected to Earth. This meant either a fluffy, snowy surface, or some kind of haze or cloud suspended in the atmosphere. A dark haze high in the atmosphere would also explain the observation of Titan made in 1971 with the ultraviolet telescope on the 2nd Orbiting Astronomical Observatory (OAO-2) by the young researcher John Caldwell. This was the first observation of any planetary satellite (other than our own Moon) to be made from an orbiting space observatory. A thick, clear atmosphere would be bright at blue wavelengths for the same reason that Earth's sky is blue – because small gas molecules scatter blue light more efficiently than red. Yet Titan was very dark in the blue and ultraviolet part of the spectrum. The logical conclusion was that some kind of dust or haze must be absorbing the blue and ultraviolet light. Otherwise it would be scattered by the atmosphere.

On the other hand, this scenario left some questions – one being what kept the haze up. If the atmosphere were thin, the haze particles would drizzle down too quickly. In the words of Ed Danielson recorded in the proceedings of the 1973 Titan workshop, 'They'd go plop'. *Pioneer 11*'s encounter with Saturn did not resolve the issue, nor was its modest complement of instruments expected to do so. *Pioneer* acted as a pathfinder for the more formidable *Voyager* spacecraft, following hot on its heels.

The two *Voyager* spacecraft, each almost a tonne in mass with over 110 kg of instruments, were launched in 1977 and had already returned an enormous volume of data about Jupiter and its satellites. Because Titan was a target of such importance, *Voyager 1* was aimed to speed by within 4400 km of its surface. This decision ruled out future planetary encounters for *Voyager 1*, which was flung out of the plane of the solar system as a consequence, but the sacrifice was felt to be worthwhile. *Voyager 1*'s meeting with the saturnian system in November 1980 was eagerly anticipated, not least because of the mystery of Titan, with the prospect of organic chemistry and a warmish surface.

About this time, a newly completed radio observatory, the Very Large Array (VLA), was used by radio astronomer Walter Jaffe, teamed with John Caldwell and Toby Owen at the State University of New York (SUNY) at Stony Brook, to measure the heat coming off Titan as microwaves. These ultra-short radio waves go clean through the atmosphere, so there would be little doubt that this measurement was probing the surface temperature with no complications from stratospheric emission, unlike the infrared measurements made previously. A narrow beamwidth is needed to ensure that only Titan's heat radiation is captured, with no contamination from nearby Saturn. Only with the VLA, consisting of many dish antennae networked together, could this be done properly. The answer was 87 K, far colder than the balmy surface some optimists had predicted. Since this measurement was made before the *Voyager* encounter, but only published shortly afterwards, in December 1980, it perhaps didn't receive the attention it might otherwise have done.

The *Voyager 1* encounter with Titan presented an opportunity to resolve the mystery surrounding the measurements of Titan's temperature. It was designed so that there would first be a good view of the dayside. Then afterwards, *Voyager* would go behind Titan, so that its radio signal to Earth would pass through Titan's atmosphere. Before the signal was blocked by Titan's surface, it would be refracted in the atmosphere, which would act rather like a weak lens. Measurements from Earth of the degree of refraction would determine how the density of the atmosphere changes with height. Making some reasonable assumptions about composition would also allow a temperature profile to be constructed. Which of the two theories to explain the puzzling results on Titan's temperature would be correct? Was the stratosphere warm, or was there a warm surface beneath a deep atmosphere?

It turned out that both theories were right – there was high haze absorbing sunlight *and* the atmosphere was deep, which allowed the atmosphere to be warmer near the surface than higher up. The reality

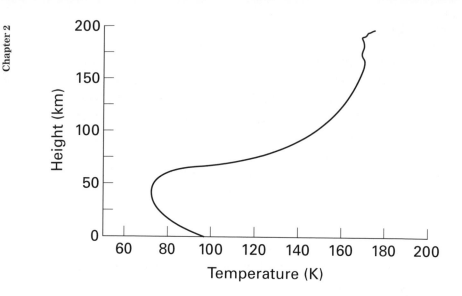

Figure 2.2. The variation of temperature with height in Titan's lower atmosphere as deduced from the *Voyager 1* radio occultation experiment.

was much as foreseen by the Canadian-born astronomer and atmospheric scientist Donald Hunten, who had organised the workshop on Titan in 1973. The uppermost part of the atmosphere – the stratosphere – was warm, as the infrared observations had suggested, because the 'dust' was absorbing short-wavelength (i.e. blue) light and, near the surface, the lower troposphere was warm because of a greenhouse effect. Hunten's predictions showed the tropospheric temperatures rising the deeper you went but, until the *Voyager* measurement, there was no way of knowing exactly how far down you could go before striking the surface. Hunten's graph had shown a dotted line trailing off into the unknown depths.

But now, post-*Voyager*, we knew. The line stopped at a pressure of 1.5 bar – 50% higher than the pressure at Earth's surface – and at a temperature of a mere 94 K. It was far colder than the muggy oasis of life that some had hoped for. But this temperature was interesting in another way, which we'll discuss more in later chapters. It is close to the 'triple point' of methane. The triple point of a substance is the temperature at which all three states – solid, liquid and gas – can coexist. Earth's interesting weather, and in particular its rain, rivers and oceans, are phenomena that occur because Earth is close to water's triple point of 273 K. Titan might not be an oasis – certainly no palm trees – but maybe it would have seas of a kind.

Picture this

So what was down there? At the time of the *Voyager* encounters, no
features on Titan could be reliably seen from Earth. That would have
to wait until the 1990s. There had been two crude pictures of Titan
made by stringing together the spin-scan data from *Pioneer 11*. They
showed Titan's apparent size and gave the first reliable indication of
how brightness varied across its disk. They also showed that, at blue
wavelengths, one hemisphere was rather brighter than the other,
although this finding wasn't fully explored until some years later. No
real features were seen in the two images. They were, after all, taken
from some 360 000 km away – almost as far as the Moon is from Earth.
Titan's disk was only about 15 pixels across.

The *Voyagers* would be a wholly different story. Each spacecraft
carried two vidicon cameras, based on the same technology as old TV
cameras, one with a wide-angle lens, the other with a narrower,
higher-magnification view. The cameras were sensitive to near ultraviolet radiation (300 nm wavelength, or 0.3 microns), through blue
light (450 nm) and green (540 nm) to orange (650 nm). They were not
sensitive to red and near-infrared light but this had not prevented
them from returning stunning vistas of Jupiter's satellites. Taken
from only a few thousand kilometres, the images of Titan should show
details less than a kilometre across. All in all, over 200 images were
taken of Titan by the two *Voyagers*.

But, much to the disappointment of geologists (and even meteorologists), most of these images were virtually featureless. Titan was
covered in a thick haze, in which there appeared to be no gaps. There
was some faint band structure and a 'detached' haze layer. The northern hemisphere seemed darker than the other and there was a dark
ring around the north pole. And that was it; no impact craters, no
cliffs, no volcanoes. Whatever mysteries lay beneath the haze were
invisible to *Voyager*'s cameras.

Of the 200-odd pictures, only a handful are ever shown in books
and articles (although the rest are archived). The classic picture of
Titan is a combination of images in different colours, which shows
Titan as a fuzzy orange ball, with one hemisphere darker and redder
than the other, and with a dark polar hood (Figure 2.3 and Colour
Plate 4). Another popular image, one of the closer ones, is of Titan's
edge or 'limb'. This shows a distinct haze layer above the main
orange haze (Figure 2.4 and Colour Plate 5). A third familiar image

was taken by *Voyager 2*. *Voyager 2* arrived at Saturn about nine months after its sibling. Its trajectory in the Saturn system was dictated by the need to get a gravity assist (the so-called 'slingshot effect') from Saturn, so the craft would travel on to Uranus and Neptune. Consequently, it didn't come anywhere near Titan, although it did take a few pictures to look for seasonal changes. These show a change of a few per cent at most in the brightness difference between the north and south hemispheres. *Voyager 2* also took some pictures of Titan as a thin crescent. Scattering by haze meant that light could be seen all the way around Titan's disk, forming a ring (Figure 2.5).

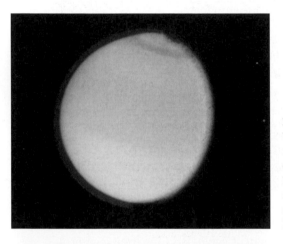

Figure 2.3. An image of Titan taken by *Voyager 2* on the 23rd of August 1981 from a distance of 2.3 million km (1.4 million miles). The contrast has been enhanced to reveal a lighter area in the southern hemisphere and a dark collar around the north pole. NASA image. (In colour as Plate 4.)

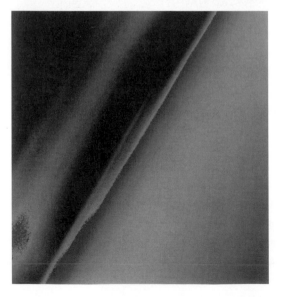

Figure 2.4. Detached layers of haze in Titan's atmosphere imaged by *Voyager 1* on the 12th of November 1980 from a distance of 22 000 km (13 700 miles). NASA image. (In colour as Plate 5.)

Figure 2.5. A *Voyager 2* image of the night-side of Titan taken on the 25th of August 1981 from a distance of 0.9 million km (560 000 miles). The atmosphere can be seen illuminated around the whole of Titan's disk. (The pattern of small black dots was built in to the vidicon camera as a means of properly correcting any distortion of images.) NASA image.

While the *Voyager* encounter with Titan was a dismal let-down for geologists, chemists were presented with a remarkably interesting object: ultraviolet and far infrared observations revealed many details about the composition of the atmosphere. We will return to them in the next chapter.

Painting by numbers

It took some years for the *Voyager* data to be even partly digested but, by the late 1980s, a reasonably clear idea emerged of how the haze distribution in the atmosphere gave rise to the observed temperature structure. Part of the problem was that the haze layer had to be very thick to explain the *Voyager* images but somehow enough sunlight had to get down to the surface to keep it as warm as it obviously was. Bob Samuelson of NASA's Goddard Spaceflight Center worked out that the haze must have properties that somehow allow red and near-infrared light down to the surface. He set his ideas down in a paper published in 1983. The theory was supported by a forbidding amount of algebra. While this approach pointed the way, further progress had to wait for some serious computing. In retrospect, perhaps someone should have realised that, if any light got down to the surface, some of it could get back up again.

NASA scientist Chris McKay, later famous for his research on exobiology, Mars and terraforming, devoted considerable effort to Titan. He worked (as did many others over the years) with Jim Pollack until Jim's untimely death in 1994. Pollack was a talented and prolific planetary scientist – Carl Sagan's first Ph.D. student. One of the projects undertaken by McKay and Pollack was a detailed analysis of the manner in which radiation travels through Titan's atmosphere. Their computer simulation broke sunlight up into two dozen or so wavelength bands, each absorbed to one extent or another by methane in the atmosphere. The atmosphere was considered as 30 layers, each with a certain density of haze particles. The optical properties of the haze particles were based on the laboratory measurements of 'tholins', complex organic substances created by Bishun Khare, Carl Sagan and others at Cornell University in laboratory experiments designed to simulate the chemistry in an atmosphere like Titan's. These attributes included factors such as how much light of each wavelength is absorbed, scattered back, or scattered forward. Particle size is of central importance and McKay used a physical model for the formation, growth and coagulation of particles to track their size as they slowly fall through the atmosphere.

Determining the characteristics of the atmosphere becomes a balancing act, in which the temperature profile through the atmosphere has to be set to ensure that the sunlight absorbed and the thermal radiation lost by each layer of atmosphere exactly compensate for each other. Any other scenario would be hopelessly unstable.

Trying to characterise with numbers what an atmosphere is like and what happens to radiation travelling through it is a complex business. Although some things about the atmosphere can be measured, many other quantities are unknown and have to be filled in with a bit of shrewd guesswork. McKay designed his simulation so that real physical quantities (such as how many haze particles are produced, how reflective the surface is and how much methane is in the atmosphere) could be varied until the predicted values of certain properties matched actual observations as closely as possible. In this way he narrowed down the range of properties the haze might reasonably have. The whole exercise gave McKay an insight into what controls the temperature balance on Titan.

McKay's model was then used by a number of other research groups. McKay himself explored the roles of the greenhouse and anti-greenhouse effects on Titan. The greenhouse effect, which raises the temperature of the atmosphere, is due to methane, nitrogen and hydrogen absorbing infrared radiation. The cooling anti-greenhouse effect is the result of haze absorbing sunlight before it can reach the surface and lower atmosphere. He found that the greenhouse effect raises the surface temperature by about 21 K but that this is offset by a drop of 9 K due to the anti-greenhouse effect. We'll discuss Titan's climate at greater length in the next couple of chapters. For now, the crucial conclusion of this research was that Titan's surface was visible – as long as one looked at the right sort of light. It showed that about 10% of the sunlight falling on Titan reaches the surface, all of it red and near-infrared. Furthermore, the brightness of Titan seen through the telescope at these wavelengths depends on how reflective the surface is. In other words, the surface could be seen. It was just bad luck that *Voyager*'s cameras were incapable of seeing at these long wavelengths.

Several research teams independently reached the conclusion in the late 1980s and early 1990s that the haze should be fairly transparent in the near-infrared. Technically, this is because the small haze particles are less effective at blocking longer-wavelength light and because the particles are themselves brighter: it is not as difficult to look through a cloud of fog as a cloud of smoke. This conclusion was becoming of more than theoretical importance as serious plans were developed for the next stage of spacecraft exploration of Titan – how best to design a camera to see Titan's surface? Unwittingly, Titan's surface had already been seen . . .

Th'inconstant Moon

While *Voyager 1* was wending its way from Jupiter to Saturn, back on Earth Dale Cruikshank and Jeffrey Morgan at the University of Hawaii searched for signs of regular variation in Titan's infrared brightness. Their results were published in early 1980, just months before *Voyager*'s arrival. They had detected what looked like a small change at 1.3 and 2 microns over a 32-day cycle. This was puzzling. Titan takes very nearly 16 days to orbit Saturn and, with the same side always facing Saturn, any effects due to surface features would be expected to produce variations with a 16-day period – only half of the 32-day one they found. Maybe the variation was in Titan's cloud layers, they suggested. Their detection, they had to admit, was some-what tentative. Even when Titan returned to the same position in its orbit it did not appear to be equally bright on each occasion. Their 18 data points were taken at irregular intervals in just over two months and were assembled from observations at three different telescopes. Comparing the measurements was difficult. Had their observations been taken on a slightly different set of dates over those two months a 16-day periodic variation might have been more obvious. They were so close!

Keith Noll of the Space Telescope Science Institute and Roger Knacke of Pennsylvania State University re-plotted Cruikshank and Morgan's data in 1993, together with a little new data of their own taken with NASA's Infrared Telescope Facility (IRTF) at the Mauna Kea Observatory, perched on the giant volcano in Hawaii and home to some of the world's largest telescopes. Their plot of Titan's brightness against orbital position showed a reasonably convincing lightcurve. Titan was brighter at some parts of its orbit than others and the pattern was constant from orbit to orbit. However, Noll and Knacke were not going to stick their necks out with speculative interpretation. 'Lower albedos appear in the western half of Titan's orbit, a property that may be intrinsic to Titan' was their somewhat cautious comment.

Around the same time, Mark Lemmon and Erich Karkoschka, Ph.D. students of Marty Tomasko at the University of Arizona, had been taking spectra at Kitt Peak National Observatory in Arizona. Figure 2.6 shows an example. They found that Titan's brightness at wavelengths corresponding to absorption bands of methane was con-stant with time. This particular radiation is being reflected by some-thing above the methane. A haze layer is just about the only candidate. Meanwhile, the brightness at wavelengths *between* the methane bands varied over the course of Titan's orbit. Titan's leading hemisphere, which is seen at greatest eastern elongation, was brighter at 0.94 and 1.07 microns (940 and 1070 nm) than the opposite hemisphere, which faced Earth eight days later.

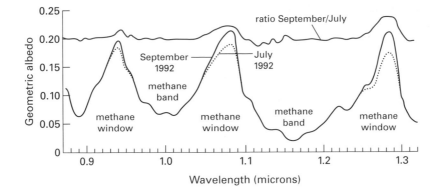

Figure 2.6. Two infrared spectra of Titan, taken at opposite sides of its orbit by Mark Lemmon with Erich Karkoschka and Marty Tomasko, provide evidence of a variegated surface on Titan. One spectrum is shown as a solid line, the other as a broken line. The brightness in the methane absorption bands, where light is reflected by the haze, was the same on both occasions. In the 'windows' between methane bands (see Fig. 2.8), where light is received from the surface, a difference in brightness hints at the presence of bright and dark regions. Adapted from a diagram by Lemmon *et al.* in *Icarus*, vol. **113**, p. 27 (1995).

Observed at wavelengths between the methane absorption bands, Titan's atmosphere is relatively transparent and one can see through it down to the surface, except that the haze uniformly distributed around Titan muddies the view somewhat. Nevertheless, theory suggests that some of the signal between the methane bands must be radiation reflected by the surface.

The clear implication was that part of Titan's surface is brighter than the rest. Technologically, this discovery could have been made 20 years earlier. Half the skill in science can be knowing the right questions to ask.

More and more data was accumulated, enabling Lemmon by 1994 to build up a good, consistent lightcurve (shown in Figure 2.7). It could be explained simply by the presence of two bright circular spots on the surface. Equally, an infinity of bright and dark surface patterns could give the same result. There was no way to tell the difference. However, the curve was approximately the same at all the wavelengths observed: 0.94, 1.07, 1.28, 1.58 and 2.00 microns.

Other research groups soon published results on the near-infrared transparency of Titan's atmosphere. They included Toby Owen and his student Caitlin Griffith, both then at SUNY Stony Brook and a team led by the Greek astronomer Athena Coustenis at the Observatory of Paris. Using the Canada–France–Hawaii telescope on

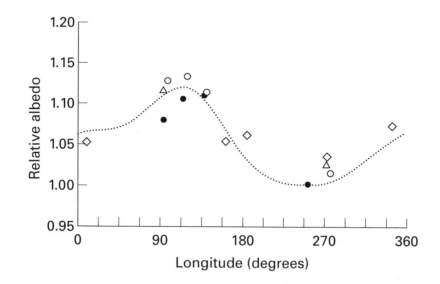

Figure 2.7. Titan's brightness variation as it rotates, measured at a wavelength of 1.1 microns. The data was assembled by Mark Lemmon and colleagues. Different symbols represent data points from various observers. There is a noticeable peak around longitude 100°. Adapted from a diagram by Lemmon *et al.* in *Icarus*, vol. **113**, p. 27 (1995).

Mauna Kea, Coustenis and her colleagues added to the collection of data points. Coustenis's group attempted to interpret the brightness at different wavelengths as evidence for particular surface compositions, using McKay's model to try and remove the effects of the atmosphere. These attempts were regarded by some as premature as there were too many unknowns. Intercomparison between each groups' data was difficult, in that they all used different instruments, different calibration stars, and so on. But the basic shape of the lightcurve was consistent whoever plotted it. Something was down there.

Eye in the sky

Spectra and lightcurves are all very well but images are what people like to see. Now that an alliance of spectra and theoretical models had convinced researchers that surface features were present, the grand challenge of seeing them seemed tantalisingly within reach.

The fundamental limitation operating against imaging Titan from Earth is resolution. With several 4-m telescopes available and capable in principle of achieving resolutions of a twentieth of Titan's diameter, lack of glassware as such was not the problem. The obstacle as ever remained Earth's turbulent atmosphere. The first way this problem was circumvented was by putting the telescope above the atmosphere.

The Hubble Space Telescope (HST), launched in 1990, provided the best images of Titan obtained in the 1990s. In the 1970s and 1980s during HST's development John Caldwell had been one of the leading advocates of using it for planetary astronomy and it was his team that obtained HST's first Titan images, in 1990. At that time, not only was the image blurred by the well-publicised defect in Hubble's main mirror but its ability to track solar system targets, which move against the fixed background of stars, had not yet been developed. The exposures had to be very short to prevent motion blur. All this meant that the quality of those three early HST images was quite poor but they yielded some important information nevertheless.

The blue and green images showed that the north–south asymmetry seen by *Voyager* had reversed: the northern hemisphere was now the brighter, as might be expected from a seasonal effect. The third image was taken in the near infrared methane absorption band, at 889 nm (0.89 microns). The quality of this image was very poor: the short exposure was really inadequate for the low intensity of reflected sunlight at this wavelength. However, it showed that the asymmetry was reversed at long wavelengths. Where Titan was bright in the blue, it was dark in the near-infrared, and vice versa.

The tracking capability was added soon afterwards and Peter Smith obtained a good set of images of Titan in 1992. These were still blurred by the bad mirror but benefited from the much better signal-to-noise achievable with longer exposures.

Two of the images were at near-infrared wavelengths. One was in a deep methane band at 889 nm. The other was through a broad-band filter known officially as the F850LP (LP for long pass), which let through radiation longwards of 850 nm, although the detector only worked to about 1050 nm, so providing a cut-off. This broad band spanned both the 889-nm methane band and the adjacent 'window' through the atmosphere at 940 nm. It was reasoned that subtracting the 889-nm image from the broad-band image should, to a rough approximation, leave an image similar to what would be obtained with a 940-nm filter. The result was indeed intriguing. The southern hemisphere was quite bright, due to scattering from the haze, while the northern hemisphere showed a couple of bright spots. This residual image was shown at *Cassini* mission planning meetings and at

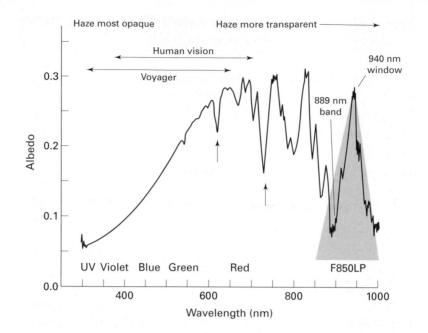

Figure 2.8. The spectrum of Titan in the visible and near infrared regions, shown as graph of albedo (the proportion of sunlight reflected) against wavelength. The general level of the curve rises towards the red end of the spectrum (at the right) because of the reddish hue of Titan's haze. Troughs occur in the curve where methane gas absorbs light. These methane bands become progressively stronger towards longer wavelengths.

The vertical arrows indicate the methane bands identified by Kuiper in his 1948 spectra (see Fig. 1.7), which led him to deduce that Titan has an atmosphere.

The haze is most transparent to radiation in certain wavelength regions, called 'windows', at the infrared end of the spectrum. These spectral windows allow telescopes to sense light reflected from Titan's surface. The most prominent window is at 940 nm, which is close to the wavelength typically used by TV remote controllers.

The shaded wedge shows the transmission properties of one of the filters used on the Hubble Space Telescope, known as F850LP. It covers both the strong methane band at 889 nm and the window at 940 nm but, since Titan is so dark at 889 nm, most of the light received when Titan is observed through this filter comes from the window region. The cameras on *Cassini* operate over the whole spectral range shown here in contrast with the *Voyager* cameras, which were not sensitive to the infrared. The wavelength ranges visible to the human eye and to the *Voyager* cameras are indicated.

conferences but was never published in a scientific journal. One reason was that it was difficult to interpret: the bright spot could be due to clouds, or surface features. With only one image it was impossible to tell. Another perhaps psychological reason was that the image looked like a smiley face, so was a little difficult to take seriously. However, the fact that details could be seen was encouraging.

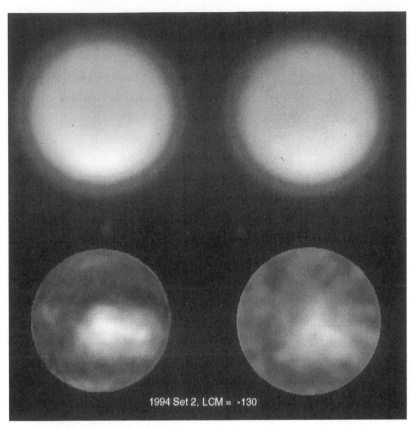

1994 Set 2, LCM = -130

Figure 2.9. Combining Hubble Space Telescope images of Titan to reveal surface features.

Top left: A single HST image through the F850LP filter (see Fig. 2.8), which probes the 0.94-micron 'window' in Titan's atmosphere. The image has been processed so that individual pixels are not visible.

Top right: A generalised 'atmosphere' image made by averaging raw images from all longitudes.

Bottom right: The 'difference' image obtained by subtracting the 'average atmosphere' from the individual frame at top left.

Bottom left: The result of averaging a set of 'difference' images into a map and projecting the map onto a sphere. All views are centred on longitude 130° W.

Courtesy Mark Lemmon.

With this promising early result, Smith was awarded telescope time to obtain a new set of images in October 1994, to map the surface and to look for the movement of clouds. This time Hubble had a new camera, which corrected the blurring effect of the misshapen mirror. These new images would be considerably sharper, with a resolution of better than a tenth of an arc second. Titan would be about 18 pixels across. A set of images was taken through visible light filters to look for seasonal change together with a set of images through the 850LP filter. Observations were made on 14 occasions. Seven of the 14 were taken in rapid succession about four hours apart the other seven at roughly two-day intervals, to obtain images throughout Titan's 16-day orbit.

RALPH'S LOG. 1994.

I had nearly finished my Ph.D. in Canterbury and had arranged to spend a year as a postdoctoral researcher in Arizona, working jointly with Peter Smith and with Jonathan Lunine (whom we'll meet later). With Peter I would be working on these exciting new images of Titan and he invited me over to a planning meeting in May 1994 at the Space Telescope Science Institute in Baltimore to get my perspective on what we might learn about the surface from the images.

I was actually moderately downbeat and showed some pictures of the Galilean satellites taken by Pioneer 10 (the twin of Pioneer 11), which also had about 20 pixels across each moon. Yes, there were bright bits and dark bits, but that didn't mean we'd be able to understand them . . .

Mark Lemmon, working with Smith, developed a computer program to convert the HST images into a map, subtracting 889-nm images to bring up the surface features, as they had done with the one pair of much poorer images earlier. This turned out not to work too well in this case. The degree of darkening around the circumference of Titan's fuzzy disk was different between the two filters resulting in a suspicious bright ring of untruth. Lemmon tried instead making an average of all the 850LP images and subtracting that from each one in turn. Adding all longitudes together would smooth out any effects due to surface features but leave the average light due to the haze. This approach worked well, although it did force all latitudes to have the same average reflectivity. However, the map that came out of the exercise was compelling: it showed a large bright region at just the longitude where the lightcurves of Lemmon, Griffith, Noll, Coustenis, and their respective colleagues, peaked in brightness. Several other smaller, fainter, bright spots were present, along with some dark regions. Lemmon worked furiously to get these maps together quickly – but it was worth it. Smith presented them at a meeting in October 1994 to no less than a standing ovation and they subsequently appeared in many magazines and books (see Figure 2.10 and Colour Plate 6).

Figure 2.10. Four views of a global projection of the map of Titan's surface assembled from 14 images taken with the Hubble Space Telescope through the F850LP filter (see Fig. 2.8) between the 4th and 18th of October 1994. The upper left image is of the Saturn-facing hemisphere. Each subsequent image (working from upper left to lower right) represents a rotation of 90°. Thus, the upper right image is the 'leading' hemisphere, the lower left is the 'anti-Saturn' hemisphere and the lower right is the 'trailing' hemisphere. The resolution is about 580 km (360 miles).

The solid shading around the poles shows areas that could not be imaged through the haze. The gap in coverage extends to the equator at about 10° longitude (top left image). The shading here shows the average intensity. The overall contrast amounts to only about 10% of the total light collected through the filter.

Credit: Peter H. Smith and NASA. (In colour as Plate 6.)

Several months later, Smith, Lemmon and Lorenz wrote up the results, in collaboration with Caldwell and two others, Larry Sromovsky of the University of Wisconsin and Mike Allison of the Goddard Institute for Space Studies in New York, who had been on the team to study cloud motions in the images. As it turned out, there were no *obvious* cloud motions to be seen. The paper, eventually published in the journal *Icarus* in 1996, included some revised maps, made by studying the brightness of each point as it rotated from the centre of the disk to the edge. This was to be the only *map* of Titan published for around four years.

In 1997 a group at the University of Hawaii used HST's new Near Infrared Camera and Multi-Object Spectrometer (NICMOS) to make a set of maps, finally published in 2000, in the 1.07, 1.6 and 2.04 micron windows. Comparison of these maps with the original set by the Arizona group is still underway as we write, but they certainly compare well, showing a number of consistently dark areas and the now-famous 'bright feature'.

Do not twinkle, little star – then we'll know just what you are

The launch of the Hubble Space Telescope in 1990 finally gave would-be observers of Titan the break-through in clarity of vision they crucially needed. But a series of set-backs had already delayed the HST launch by several years and after it was finally in orbit, astronomers rapidly learned to their dismay that the main mirror was misshapen and its images defective. The problem wasn't finally fixed until the end of 1993. Meanwhile, back on the ground, technological developments were being perfected that would allow telescopes below the atmosphere to compete on image quality with the HST circling far above them and ultimately surpass the performance of HST in certain circumstances.

With a main mirror 2.4 m (94 inches) across, the HST's light-gathering capacity and power of resolution is modest. Even when it was launched, the largest ground-based telescopes were in the 4-m class giving them two to three times the HST's light collecting area and double the resolving power. The next generation of large telescopes were already being planned and constructed, with mirrors of 8 and 10 m. HST's unique advantage was its location, beyond the tiresome teasing of the atmosphere. But that advantage was about to be challenged. Ground-based observatories wheeled out the first of their secret weapons – adaptive optics.

The development of adaptive optics was stimulated in part by the USA's Strategic Defense Initiative, popularly known as 'Star Wars'. The idea was to send powerful laser beams through the atmosphere to destroy satellites or incoming missiles, either directly or by bouncing them from mirrors in space. The turbulence in the atmosphere would bend and distort the beams in the same way that it affects astronomical images but, if the mirror could vary its shape very rapidly, it could be made to cancel out the atmospheric distortion. This is technologically difficult. There has to be a way of accurately sensing the distortion caused by the atmosphere and rapidly feeding back the information to actuators on the mirror, which respond by changing the mirror's shape appropriately. The advanced computing and actuators required became available to astronomers in the early 1990s. In an astronomical adaptive optics system, a small thin mirror that can easily be deformed by actuators on its back is placed in the path of the light that has been collected by the telescope. Atmospheric disturbances are monitored by the image of a bright reference star or by

Figure 2.11. A 2-micron image of Titan taken in 1999 with the 10-m Keck II telescope, employing adaptive optics. It was obtained by a team led by Claire Max. North is towards the lower left. Bright haze can be seen all around the disk but it is brightest in the south (upper right). Bright and dark surface features are obvious; the contrast in this image is excellent. The dark feature near the centre is suggestive of a large lake, perhaps an impact crater basin.

using an 'artificial star' produced by laser light projected through the telescope into the atmosphere.

Distortion caused by the atmosphere gets worse at shorter wavelengths. Put another way, less turbulence is needed to bend a blue image than a red one by some specified amount. Because of this, the first astronomical images using the adaptive optics technique were made at the longest practicable wavelength – 2 microns. More recent work has successfully produced images in the 'windows' through Titan's atmosphere at 1.6 and 1.3 microns but it gets progressively harder to do at shorter wavelengths.

The first image of Titan by this technique was made by a French-led group and published in 1993. They imaged the asteroid Pallas as well as Titan and found that both objects were distinctly non-round. They suggested that there are spots on Titan but, since they had only one image, it wasn't possible to discriminate between clouds and surface features. Athena Coustenis and co-workers took a number of adaptive-optics images of Titan in the mid-to-late 1990s with the European Southern Observatory's 3.6-m New Technology Telescope in Chile. These reassuringly showed the same shape of bright region as the HST map. The 1994 HST data had hinted at the possibility that the bright region was not uniform. These ground-based images were at wavelengths that penetrated the haze more easily and they showed more clearly that there are three brightness 'peaks'. Perhaps these were mountaintops. However, because the observations were not distributed evenly in longitude around Titan, there was not enough information to make up a map.

Along with adaptive optics, a second line of attack was also successfully implemented to combat the atmosphere. Called speckle imaging, it had been demonstrated as a viable method a couple of decades earlier but it only came into its own in the 1990s. Speckle imaging relies on the fact that, while turbulence makes an image dance around, it doesn't destroy it. If a camera is used in the conventional way, with the shutter opened for several seconds or minutes, the dancing image is smeared and the detail is lost whether the detector is electronic or photographic film. However, if the shutter is opened only briefly, say for a fraction of a second, the dancing image is frozen. The only difficulty is that the shorter the exposure, the noisier the image. Simply adding the images together reproduces the effect of a long exposure, which improves the signal-to-noise ratio but blurs the image. But if a stack of such images can be shifted correctly with the aid of a computer, so that they all fall exactly on each other, the noise level is drastically reduced without the image being blurred. In essence, speckle imaging does electronically what adaptive optics does mechanically.

The speckle technique is implemented, somewhat imperfectly, by the human eye and its formidable image processor, the brain. This accounts for the remarkable quality of some of the naked-eye observations by Comas Solà, Dollfus and others. Another way it can be done is with a video camera. Using an ordinary video camera attached to a small telescope, amateur astronomers at the Boston Museum of Science have made striking pictures of Earth-orbiting satellites, like the Mir space station. The video exposures of a fraction of second are short enough to remove most blurring. The processing demanded considerable patience as it involved looking at thousands of individual frames, selecting the best ones and adding them. But the results are striking: details like the solar panels, or the docked Space Shuttle can be seen. Such a labour-intensive procedure, and one reliant on subjective evaluation of the images, is not appropriate for professional, quantitative observations. Instead, images are broken down into frequency components, a process that lends itself to automation rather better.

Speckle imaging takes immense computing power and, because the exposures have to be short, it only works for bright objects. Remarkably, the first such observations of Titan were made in 1981, just a year after the *Voyager* encounter. They didn't tell us much more than we knew already but at least they demonstrated the technique. In

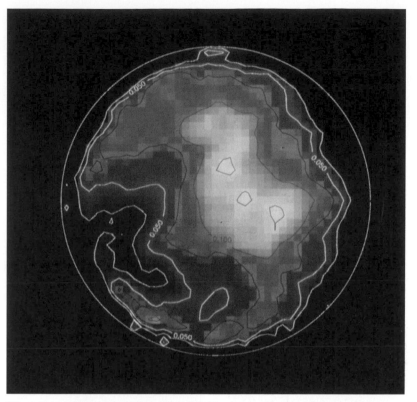

Figure 2.12. The surface of Titan's 'bright' side recovered by means of the speckle imaging technique by Seran Gibbard and colleagues from images made with one of the 10-m Keck telescopes. This observation made the significant discovery that the dark region to the lower left is literally pitch black, suggesting perhaps liquid hydrocarbons on the surface. Image courtesy Seran Gibbard.

1996, a group led by Seran Gibbard at the Lawrence Livermore Laboratory in California, using the 10-m Keck I telescope, obtained excellent quality 2-micron images with a resolution rather better than those taken by Hubble. They used McKay's computer model of the atmosphere to remove the effects of scattered light and estimate the reflective power of the surface. Their results, published in 1999 and shown in Figure 2.12, were encouragingly similar to the HST map. They also found that the dark regions had a very low reflectivity of about 2%. Hydrocarbon seas, something that had been long suspected on Titan, would have this value but this was one of the first quantitative pieces of evidence in favour of that picturesque idea. This important new result was not ironclad proof of liquids on the surface since almost all organic materials are as dark. But it was highly suggestive.

A distant echo

During the *Voyager* encounter with Titan, it occurred to Charles Elachi, at the time a JPL engineer, that a radar would be the ideal instrument to explore Titan's hidden surface. At that time he was embroiled in the early development of a mission then called Venus Orbiting Imaging Radar (VOIR). Ultimately, about a decade later, it evolved into the *Magellan* spacecraft, which unveiled the surface of Venus hidden beneath perpetual clouds. The first spaceborne imaging radar, on a satellite called SEASAT, had flown only three years before.

In the early 1980s, Carl Sagan and dynamicist Stanley Dermott found an intriguing angle on the problem of whether Titan was wet or dry. They argued that, if there were seas of methane (about which more in the next chapters), then tides would have eroded any mountains or hills, smoothing Titan until it resembled a billiard ball covered with a uniform sea. Furthermore, they said, since Titan's orbit is elliptical to this day, the quantity of energy dissipated by any tides on Titan would have to be quite low. If tides were extracting energy from Titan's motion, the effect would have made Titan's orbit circular by now. The implication of this was that seas on Titan had to be more than 300 m or so deep, if they were there at all. Sagan and Dermott also noted that if Titan really were covered by deep ocean, it would appear dark to planetary radar. In making this remark they were anticipating the fact that Saturn's changing position in the sky would bring Titan within the scope of the largest radio dish in the world, 305 m (1000 feet) across and nestling in a crater-like depression at Arecibo in Puerto Rico. The natural formation is in fact a limestone sinkhole – not an impact or volcanic crater. The Arecibo dish can be used as a radar transmitter as well as a radio telescope and is operated by Cornell University, where Sagan and Dermott both worked at the time. Sagan and Dermott's paper, published in the prestigious journal *Nature*, was elegant and innovative. But it was too idealised and, in retrospect, irrelevant to the debate on Titan's surface. However, it shaped thinking on the subject for the following 15 years. It posed a neat observational test, one that challenges instrumentation to this day.

Radar development was enormously stimulated by the Second World War and those developments in turn helped the emerging new technique of radio astronomy. Some of the radio astronomers also experimented with radar, using it to study meteors and other phenomena. A radar echo was bounced off the Moon as early as 1946. The icy Galilean satellites of Jupiter were successfully targeted by radar in the 1970s. They turned out to have rather bizarre radar properties. First, the echo was several times stronger than that expected from experience with the Moon and the terrestrial planets. Secondly, the polarisation of the radar echo was strange.

Polarisation

Polarisation is a particular property electromagnetic radiation may or may not have. In a beam of unpolarised light, the oscillating electric and magnetic fields that propagate the radiation and give it its name are oriented equally in all possible directions at right angles to the path of the light. In polarised light, these oscillations are concentrated in one particular orientation – horizontally or vertically, for instance. If the orientation is fixed, the radiation is said to be 'plane' or 'linearly' polarised. Natural sunlight is unpolarised but reflected light typically is polarised. Most of the light reflected from the surface of a river is polarised horizontally, while the light that penetrates into the water is vertically polarised. If you view a river at the correct angle through polarising sunglasses, much of the light reflected from the surface of the water can be blocked, while everything under the water – the fish swimming in it for example – appears much clearer.

As well as linear polarisation, there is another variety. Instead of being fixed, the orientation of the electromagnetic fields can rotate at a steady rate as the beam of light travels. The tip of an arrow representing the direction of the field would follow a track shaped like the thread of a screw. This phenomenon is called 'circular' polarisation. Like screw threads, it may be 'left-handed' or 'right-handed', according to the direction of rotation.

Radar astronomers choose to use circularly polarised radiation because its polarisation properties are not altered during its passage through space and Earth's ionosphere, whereas linearly polarised radiation is affected. On reflection from a smooth solid surface, the radar beam's sense of polarisation is reversed – a right-hand circular signal will switch and become left-hand polarised. The radio receivers on the ground can be oriented to be sensitive to one polarisation or the other.

This reverse-sense polarisation was the kind of echo that came back from the Moon and terrestrial planets. Yet the echo from the cold, icy, Galilean satellites was predominantly of the same sense as the transmitted signal. One idea was that the radio waves were scattered inside the icy surfaces. Bouncing many times, the polarisation would be switched around a random number of times – often an even number – so that the polarisation would finally be flipped back to its original sense. This idea relied on ice being very transparent to radio waves. As it happens, at very low temperatures it is.

Astronomers tried on many occasions to get an echo from Titan. The first reported successful measurement, and the only observation at the time known to contain information about Titan's surface, was reported in 1990 by Duane ('Dewey') Muhleman of the California Institute of Technology ('Caltech') in Pasadena and his colleagues. They had used the powerful Goldstone 64-m radio telescope in California to transmit towards the Saturn system and the sensitive Very Large Array in New Mexico as the receiver. The signal took about two and a half hours to come back.

Even after hours of integrating, the signal level was only about double the noise. The reflectivity they found was unexpectedly high at 15–30%. This was a real challenge to the theorists. Titan's surface looked icy to radar but how could this be if the surface was covered by deep hydrocarbon seas that chemical models suggested should be there? And if it wasn't totally covered in deep hydrocarbon seas (remember this was four years before the HST map), it was a puzzle how the orbit had managed to stay elliptical according to the argument of Sagan and Dermott. It was a paradox that promoted some exotic ideas. One that still endures is that Titan may have a porous surface, so liquids could fill pores, cracks and caverns beneath the icy surface. Other ideas, now abandoned, were that the seas were frothy or laden with ice particles. Even though ice should sink in hydrocarbons, the desperate theorists, literally clutching at straws, advocated stringy particles forming floating mats.

Muhleman and his colleagues also reported that there was a bright spot, at the same longitude as the bright spot later seen in infrared lightcurves. One of their three nights of observations gave a higher return than the others. Furthermore, when they repeated their observations the following year, they found a bright spot again but one day earlier than it should have appeared. Since Titan was always expected to rotate synchronously, turning once on each orbit, this was quite a surprise. Muhleman's team suggested that Titan could have a nonsynchronous period – 15.911 days rather than the 15.945 days that would be truly synchronous. The difference between these two periods would give a difference in longitude of 1 day (1/16th of a period) over 1 year, or about 20 revolutions.

The result was greeted with some scepticism but it had been suggested on theoretical grounds by Richard Greenberg that the Galilean satellites might rotate nonsynchronously. If the orbit is not quite circular, then nonsynchronous rotation is stable. Graduate student William Sears, working on tides with Jonathan Lunine and Richard Greenberg (all at the University of Arizona), showed that the period suggested by Muhleman was indeed possible.

RALPH'S LOG. 1994.

When I completed my Ph.D. and started as a postdoctoral researcher with Jonathan Lunine and Peter Smith in Arizona in November 1994 (the same week, incidentally, that William Sears defended his Ph.D. thesis and a week or two before Lemmon finished his), one of my first jobs was to see if the new HST observations could prove or disprove the nonsynchronous rotation hypothesis by checking whether features moved from one image to the next at the expected synchronous rate. Since the images spanned only about 16 days, the relative movement, if the period were nonsynchronous, would have been only about 1°. A single pixel in the images was about 6° on Titan and picking out the fuzzy edge of the bright feature with sufficient accuracy was too difficult, given the intrinsic noise in the images. With some new images in 1995, however, the length of time between various pairs of images was enough to eliminate a nonsynchronous period. Muhleman's 1990 report had been a red herring. Lemmon also compared a Titan spectrum taken in 1979 with his own taken in 1992/3 and showed that it must have been Titan's bright side. This too was inconsistent with nonsynchronous rotation.

In 1995 Muhleman published a review of planetary radar observations. The numbers in the 1995 paper didn't agree with the earlier published numbers. Bryan Butler, who had been Muhleman's Ph.D. student at the time the 1995 paper came out, explained: it turns out that a factor of two, deeply buried in a computer program at the Very Large Array in Socorro, New Mexico, was wrongly introduced in the radar reflectivity measurements, making the originally published value too large. This factor of two was corrected in the 1995 paper, albeit without comment; the authors can't be blamed for being reluctant to draw attention to this embarrassing correction with a fanfare.

The radar reflectivity measurements made sense at last but the fact that a 'bright spot' had appeared a day early still wasn't explained. In all probability, the signal-to-noise was worse than Muhlemann believed and he over-interpreted what he saw. In retrospect, researchers hungry for any clue to the nature of Titan's surface had perhaps given too much credence to a marginal observation.

Had the correct value for the radar reflectivity been published in 1990 our perception of Titan would have been rather different. As it was, the mistaken result, together with Sagan's tidal arguments, had to a large extent 'killed' the ocean in most people's minds.

So what did the corrected lower value of reflectivity actually mean? First, it meant that Titan wasn't anything like the Galilean satellites, at least in the way it responded to radar. That was an enormous puzzle. After all, on the basis of size and mass, Titan is far more like Ganymede and Callisto than anywhere else. In the near infrared, from both spectroscopy and imaging, we knew that parts of Titan were dark, perhaps covered in organics. But other parts were moderately bright – perhaps icy – and Coustenis's spectra certainly suggested water ice in some quantity. Could a mix of organics and ice explain the discrepancy? Or was something else needed – rock, perhaps? After all, as far as radar went, the new half-as-reflective Titan looked like the Moon. But that would be stupid. That meant a coating of dense rock on top of the less dense ice that must surely make up much of Titan. Like sugar sprinkled on a cappuccino, it should sink over time and disappear.

RALPH'S LOG. 1997.

An idea I had in 1997 was that the ice on Titan might be 'different' by virtue of being tainted with ammonia. There are quite good theoretical reasons for thinking that ice on Titan might be ammonia-rich. First, while ammonia is more volatile than water, it is less volatile than methane. If Titan acquired methane when it formed (and since we see methane in its atmosphere, this seems a safe assumption) then it would have accumulated ammonia too. This ammonia probably yielded the nitrogen atmosphere we see today.

Ammonia is very soluble in water, so in the late stages of Titan's formation, when its surface was probably molten ice (i.e. liquid water), the ammonia in the atmosphere would have been in equilibrium with ammonia dissolved in the water. Eventually, the ammonia-polluted water began to freeze over.

Pure water freezes at 273 K (0 °C) but a concentrated solution of ammonia (30% by weight) will stay liquid down to a minimum freezing point of 176 K (−97 °C). The ammonia is a very effective antifreeze. The liquid on Titan's surface would have been a weak ammonia solution. As this gradually cooled, pure water ice would have formed and floated on top, while the liquid left underneath became a stronger and stronger solution.

If at some later time in Titan's history the ice cracked and liquid squeezed up through the cleft, it would have been a solution of ammonia in water at the 30% concentration at the minimum freezing point. If there are icy lava flows on Titan, then they are probably 30% ammonia.

About the time I started to think about this, I had also started working on the calibration of an instrument destined for the Martian polar terrain, so I was back in a lab, after two years office-bound. The instrument, called TEGA (Thermal Evolved Gas Analyser), was intended to cook soil samples and measure gases given off, so I was playing with chemicals and gas flows. While I was in the lab, I thought I'd try out a quick experiment on the electrical properties of ammonia-doped ice.

Concentrated ammonia solution has an overpowering smell. An uncovered beaker of it will make your eyes water in seconds so the experiments had to be done in a fume cupboard. My first experiment involved pouring ammonia solution between two metal plates, freezing with liquid nitrogen (down to 77 K, the same as −196 °C) and measuring the simplest electrical properties of the set-up – its capacitance and resistance. These basic experiments at least showed that ammonia-rich ice behaved differently from pure ice.

We think Europa reflects radar waves (i.e. microwaves) strongly because cold ice doesn't absorb the radar waves at all well and they can rattle around between cracks before being reflected back. If Titan's ice were somehow more absorbing, that would explain its weaker echo. I'd need to measure how well my ammonia-rich ice absorbed radar-type frequencies – millions of times higher than I had tested in my first simple experiment. But how? I had no strong background in radio engineering, and the measurement techniques I knew about used arcane and expensive equipment like waveguides and 'Network Analysers'.

I was thinking about this question on a flight back from the UK the following summer. It would be difficult to measure the small reduction in energy in a signal when it is transmitted through a weakly absorbing sample. But maybe the small amount of energy absorbed by the sample could be measured. That would be difficult too. Perhaps with a *sufficiently powerful source of microwaves the power absorbed by the sample would be large enough to measure reliably. But I would need a source of short-wavelength radiation with a power of hundreds of watts. Duhhh! It struck me somewhere over northern Canada: a microwave oven was just such a source. And, conveniently, there was a microwave oven in the kitchen at home . . .*

It was really rather interesting reading up in the library about microwave heating. There has been a lot of work devoted to this topic for the food industry. Just putting the sample in an oven wouldn't work. The microwaves would bounce around between the metal walls of the oven cavity until they were absorbed. It wouldn't matter how absorbing the sample actually was. The trick would be to arrange things such that the microwaves only went once through the sample.

My wife, Elizabeth Turtle, who had started studying for her Ph.D. in Arizona the same year as Mark Lemmon, tolerated my experiments with patient amusement after she had established that they weren't going to ruin our microwave oven. I preferred not to do anything bizarre in the lab until I knew it was going to work at home.

When a large beaker of water was in one corner of the oven, the microwaves tended to be absorbed after only one or two bounces. The much smaller sample would only intercept a small fraction of the energy, which would then be proportional to its absorptivity. This method seemed to work. The heat absorbed correlated very well with the known properties of samples like corn oil, plastics, kerosene (which didn't get hot at all, although the notion of putting flammable fuels in the microwave quite reasonably sounds like a bad idea). So it was time to try the experiments for real. I bought some blocks of dry ice (frozen carbon dioxide, which will chill stuff down to −78 °C, or 195 K) and used it to freeze water and dilute ammonia solutions.

After three minutes on full power, the temperature of the pure ice increased by several degrees. So far so good.

However, when I put ammonia-doped ice into the oven I was in for a shock. After only a minute or so, the ammonia-ice sample had melted, or rather partially melted – a phenomenon called thermal runaway. The same effect occurs when you put, say, frozen vegetables, in the oven. The edges warm up first and the ice in them melts. These molten regions are much more absorbing and so suck up all the microwaves, leaving you with steaming edges but still-frozen insides. This is why the 'defrost' setting on a microwave is at a lower power level than 'cook': the heat dumped into the corners has time to be conducted into the inside to give more even heating.

Doing experiments again, with bigger samples (to minimise the heat-leak effects) and heating for a shorter time, gave better results. It turned out that even with only 0.2% ammonia, ice would be made 100–1000 times more absorbing than pure ice. It all hung together if Titan's surface was ammonia-rich. Certainly this hypothesis makes some sense, although other mechanisms may be responsible for the strange radar returns from Titan. Ammonia-rich ice (despite Arthur C. Clarke's description of 'garish ammonia ice floes') is optically white, so while it is radar-dark, it could be every bit as bright visually as pure ice.

Anyway, I learnt a lot doing the experiments, which were rather fun. A number of colleagues were amused by the technique. Some others thought it was a great educational tool. Somewhat later I got a microwave oven for the lab for the specific purpose of these experiments and had a student study a broader range of ices and temperatures (once again taking advantage of TEGA's liquid nitrogen supply). Using liquid nitrogen at 77 K allowed tests on 30% ammonia solutions, which don't freeze until they are below 176 K – far colder than I was able to achieve with dry ice at home. When the sample was so cold, the problem of heat from the air became quite significant. More heat was getting into the sample from the air than was being deposited in it by the microwaves, but by going through the motions of the experiment without the microwaves on it was possible to measure this effect and to establish a reasonable absorptivity for ammonia-rich ice.

There is a sad postscript to all this, however. The TEGA experiment was lost, with the rest of the Mars Polar Lander (including a stereo camera built by our own Peter Smith), in December 1999, a reminder that space exploration is far from trivial and risk-free.

Maybe ammonia is the answer. Looking ahead, an interesting way of testing this idea will be to study the other saturnian satellites with the radar on the *Cassini* spacecraft. Are they radar-dark too? Are there radar-dark surface areas that are visually bright? Maybe it will be possible to identify ammonia in these locations spectroscopically. Unlike on Titan, these surfaces are not concealed by an opaque atmosphere.

Or maybe – as new infrared spectroscopy of Titan in 1999 and 2000 is hinting – there isn't a puzzle after all. What was thought to be a moderately bright surface is actually fairly dark. The light that we thought was coming from the surface is actually being reflected from a thin low cloud layer. If Titan is dark both optically and to radar, then maybe organic sludge can explain everything. Of course, we know from the HST and ground-based infrared work that Titan is not uniform. There are bright bits and dark bits however dark it is on average. The organics must be thinner or thicker, or different in composition, from place to place.

The early radar observations by Muhleman and others really pushed the limits of the technique. But as we write this chapter, the inexorable celestial clockwork is moving Titan into the crosshairs of the really big radar gun – the Arecibo radio dish. With its enormous collecting area and new sensitive receivers, it will achieve much better signal-to-noise than the earlier measurements. There will even be so much signal that it can be chopped up in time and frequency. Earlier echoes come from closer to the telescope – the parts of Titan closest to Earth, or the centre of Titan's disk, while later echoes come from the limb. As Titan rotates, the edge coming toward Earth is Doppler-shifted to a slightly higher frequency ('blue-shifted'), while the receding limb is red-shifted. Processing the signal this way – when there is enough signal – breaks the echo down into different parts of Titan: 'Never underestimate the power of long wavelengths' JPL radar astronomer Steve Ostro admonishes his optical colleagues. Over the next few years, Ostro and his colleagues hope to build up a radar map of Titan with about the same resolution as the HST map. What's more, while looking at Titan in the conventional manner can only show us the top fraction of a millimetre of the surface, radar offers the prospect of getting under the skin of the planet without being distracted by the atmosphere. Preliminary results are both promising and extremely tantalising according to Ostro. Titan's surface seems to have a very variable radar reflectivity and the return signal's polarisation is unlike that received either from Jupiter's icy Galilean satellites or from the rocky surfaces of Venus, Mars or the Moon.

* * *

This chapter has set the stage for the rest of the book, describing how we know what little we do about Titan's surface. It has not, you will notice, been a clear-cut, black-and-white story of hypotheses neatly formed and crisply decided by carefully orchestrated experiments. The reality of science is that it is full of false starts and dead ends. That's often why its pursuit is so fascinating.

Titan's puzzling atmosphere

How many conversations begin with a discussion on the weather? As far as human life is concerned, the tumultuous goings on in the atmosphere of our planet Earth are of immense consequence, from day to day, throughout the year and over the centuries. And we devote considerable resources to recording and predicting the ever-changing behaviour of our blanket of gas. We worry about the causes of global warming and thinning of the ozone layer. Worlds with atmospheres have an extra dimension of complexity over those that do not, an extra layer to be analysed, characterised and understood.

And so it is with Titan. Its surface may be hidden but between us and that secret landscape lies an amazing layer of gas. Can there really be a moon with such an atmosphere? It's still a novel phenomenon. What is this stuff? What is it doing there?

In Chapter 1 we described how Titan's atmosphere was properly discovered in 1944 (if you don't count Comas Solà's irreproducible 1908 report), and in Chapter 2 we related how the *Voyager* mission finally established the atmosphere's thickness. In this chapter, we'll go into more detail about what it's made of, how we think it was created (admitting right away that no-one knows for sure), and how it has evolved. Most intriguing of all is what Titan's atmosphere might be able to tell us about our own atmosphere and even, perhaps, the origin of life on Earth.

The mystery gas

Back in the mid-1970s, there were competing ideas about the unknown composition of Titan's atmosphere. After Kuiper's discovery of methane, the next substance to be identified in the atmosphere was hydrogen, found by Larry Trafton. In some respects this was no surprise, since hydrogen dominates the atmospheres of all the outer planets. On the other hand, hydrogen is a very light molecule and, as James Jeans noted in 1925, it shouldn't hang around on a body with gravity as weak as Titan's. Its existence did present a puzzle.

Careful study of spectra suggested that there was something else there besides the hydrogen and methane. Ammonia was discussed at some length as a possible candidate and, without knowing how low the surface temperature was, there was no way of ruling it out. Nitrogen was considered as an outside contender, although it was favoured by Tucson scientist Donald Hunten at the 1973 workshop on Titan.

Even in those days, researchers knew that methane would undergo chemical changes as a result of being bathed in ultraviolet light, a process called 'photochemistry'. A pioneering experiment had been undertaken by Harold Urey and Stanley Miller in the 1950s: they took a mixture of gases and water vapour, kept it warm and passed an electric spark through it for a few days. The liquid in the flask turned a little pink at first and eventually became a tarry sludge. When they analysed its contents they found many important organic molecules. Later experiments found that similar results could be obtained by exposing gas mixtures to ultraviolet light. A more recent version of this experiment, using a laser as the energy source, is illustrated in Figure 3.1 and Plate 7.

Exactly what the products of photochemistry in Titan's atmosphere would be no-one knew. 'Asphalt' was the catch-all term used by the 1973 workshop participants. Carl Sagan reported early results of experiments of his own, following Urey and Miller, in which electric sparks were passed through methane gas mixtures. This produced a reddish sludge, for which Sagan coined the term 'tholin' inspired by the Greek word for mud. Sagan argued that something of the kind could account for the orange colour of Titan. As a haze, it might be responsible for the unusual manner in which Titan reflected light and for the apparently warm stratosphere.

The revelations of *Voyager*

When *Voyager 1* arrived in the saturnian system in 1980, it was carrying an instrument that immediately revealed the identity of the mystery constituent in Titan's atmosphere. The ultraviolet spectrometer unambiguously detected the faint glow of nitrogen. There was the answer. The combination of *Voyager*'s radio occultation experiment, ultraviolet data and infrared spectra indicated that the atmosphere was between 85% and 95% nitrogen, with most of the rest methane. There was about 2% methane in the stratosphere, although near the cold surface, the amount might be nearer 6% or 8%. Argon remained a wild card and then there was one or two tenths of a per cent of hydrogen.

But the exact mix of gases remains uncertain to this day. Is there 5% of argon, or 0.5%? It sounds like a very academic question but it bears directly on such disparate concerns as the origin of the atmosphere and

(a)

(b)

(c)

Figure 3.1. Simulating lightning on Titan. In the experiment pictured here, which was carried out in 2001, a focused (invisible) laser beam was used to make a white-hot plasma in a flask of methane and nitrogen (a). After a short time, the flask became fogged with a white deposit (b). The deposit became brownish as it thickened (c). The brown deposit of organic chemicals resembles Titan's haze. Laser plasma experiments are more convenient than the electrical discharge and ultraviolet illumination experiments conducted by Miller, Sagan and others because the deposit is generated in only a few hours rather than days. Images courtesy of Rafael Navarro-González, Laboratory of Plasma Chemistry and Planetary Studies, National University of Mexico. (In colour as Plate 7.)

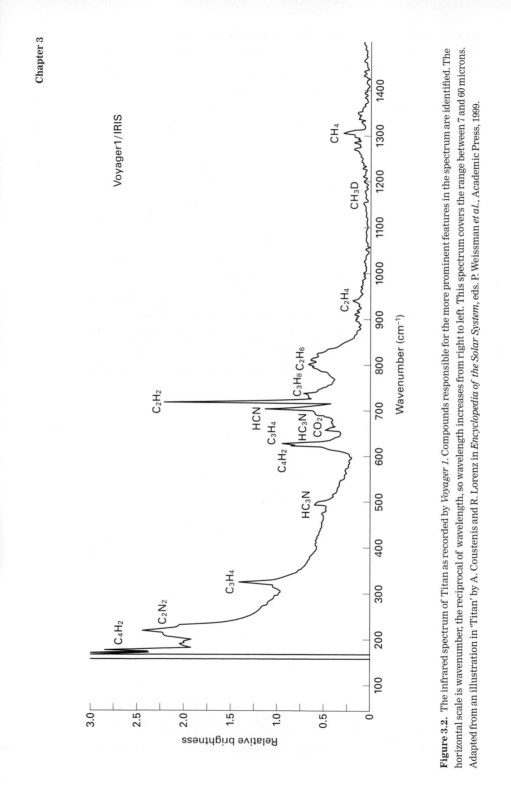

Figure 3.2. The infrared spectrum of Titan as recorded by *Voyager 1*. Compounds responsible for the more prominent features in the spectrum are identified. The horizontal scale is wavenumber; the reciprocal of wavelength, so wavelength increases from right to left. This spectrum covers the range between 7 and 60 microns. Adapted from an illustration in 'Titan' by A. Coustenis and R. Lorenz in *Encyclopedia of the Solar System*, eds. P. Weissman *et al.*, Academic Press, 1999.

the design of the heat shield of the *Huygens* probe. If the argon abundance is at the high end of the range, then the methane abundance near the surface also has to be higher, to account for the average molecular weight of the gas mix there. We know the average molecular weight of Titan's atmosphere is close to that of pure nitrogen – 28 – but argon is a heavy gas with a molecular weight of 40. If argon is plentiful, the average molecular weight is being dragged down by extra methane: methane's light molecules weigh in at only 16. Methane in Titan's atmosphere is the analogue of humidity in Earth's, so the amount is crucially important for the stability of the climate and for meteorological processes such as rainfall and clouds, which we'll go into in Chapter 4.

It took over a decade for the infrared data to be fully understood. The myriad of organic compounds took a lot of unravelling. Many effects had to be folded into the analysis – instrumental corrections, the temperature profile between the surface and the stratosphere, the angle of view. Such a detailed and painstaking investigation is usually delegated to a Ph.D. student and this one was no exception. The job fell to Athena Coustenis. By fitting the observed spectra, such as the one in Figure 3.2, to the predictions of a detailed simulation, she was able

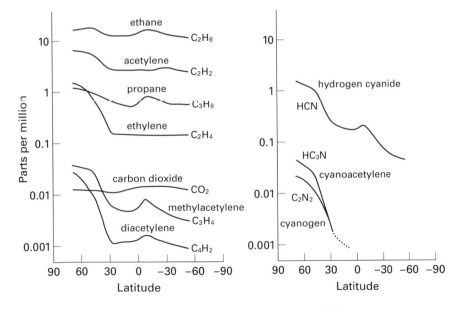

Figure 3.3. Variations with latitude on Titan of the concentrations of different gases as measured by *Voyager 1* and analysed by Athena Coustenis and Bruno Bézard. Several of the compounds are noticeably more abundant at high northern latitudes where the polar hood was observed. Adapted from a diagram by Coustenis and Bézard published in *Icarus*, vol. **115**, p. 126 (1995).

to calculate the abundance of about a dozen different kinds of molecules present in the atmosphere. Furthermore, she established that the abundances of some of them varied with latitude: ethylene and the more exotic methylacetylene and diacetylene, together with various nitriles, were around ten times more common at high northern latitudes than elsewhere. This observation could be put down to a seasonal phenomenon. The *Voyager* encounter took place during northern spring on Titan and the high northern latitudes had just emerged from a long polar night. They had been in darkness for years. The chemistry here would have been very different without light to drive photochemical reactions and without any warmth from the Sun. At the lower temperatures here, some chemical reactions would be slower than at other parts of the globe and some compounds would condense. In fact, exactly the same situation occurs on Earth. During the polar night, the atmospheric chemistry subtly alters and a rare but important gas becomes severely depleted: the now notorious ozone hole opens up.

The latitudes at which these chemical perturbations occur correspond closely with the dark polar 'hood' apparent in *Voyager*'s blue-filtered images of Titan. Dynamically this region may be a circumpolar vortex, a sort of Sargasso Sea in the sky, where the chemistry is isolated from the rest of the atmosphere by winds slavishly following lines of latitude. This is the kind of circulation pattern that occurs around the Antarctic on Earth, where the more severe ozone hole builds up. Perhaps understanding Earth's polar photochemistry will help us understand Titan's, if not vice versa.

Coustenis also studied infrared spectra from *Voyager 2*. Even though this spacecraft approached Titan no closer than 600 000 km, its spectra could be compared with those from *Voyager 1* obtained nine months earlier. Nine months out of Titan's 30-year seasonal cycle is the equivalent of only one week on Earth. Even so, there were significant variations in the abundance of methylacetylene and diacetylene at high latitudes, perhaps the effects of spring sunlight on the polar hood.

As a result of their analyses, Coustenis and her colleagues Daniel Gautier, Bruno Bézard and Emmanuel Lellouch at the Observatoire de Paris in Meudon, were able to say something about the distribution with height of several of the gases. Since most of them are produced at high altitudes and are removed by condensation lower down, their abundances decrease at lower altitudes. Such details are not of broad interest, perhaps, but they are vital when it comes to distinguishing between different chemical scenarios.

Something in the air

Methane, made of one carbon atom and four hydrogen atoms, splits when it receives a photon of ultraviolet light. It breaks up into one or two loose hydrogen atoms and a fragment containing a carbon atom and two or three hydrogens. These loose atoms and fragments (called 'radicals') are fiercely reactive. They quickly combine with whatever they find but their couplings are brief, each being followed by rapid divorce until their promiscuity is halted because all their energy is dissipated. Most of the hydrogen atoms ultimately combine with each other in pairs, forming molecular hydrogen. Light-weight hydrogen molecules move rapidly enough to escape from Titan's weak gravity and fly free into space over a period of about 1–10 million years. They do not, however, have enough energy to escape so quickly from mighty Saturn's gravity, so they collect in a doughnut shaped cloud around Saturn, forming a 'torus' enclosing Titan's orbit. The fragments of hydrocarbons combine in countless different ways to form more complicated organic compounds. For example, two methane molecules each missing a hydrogen may combine to form ethane.

Figure 3.4. A simplified schematic diagram of the principal photochemical processes taking place in Titan's atmosphere. Ultraviolet light from the Sun breaks methane (CH$_4$) into various fragments, called radicals. Symbols representing these radicals are enclosed by a dotted line in the diagram. They recombine into various organic molecules, the most abundant of which are C$_2$H$_2$ (acetylene) and C$_2$H$_6$ (ethane). They can react with other hydrocarbon radicals and nitrogen radicals from the break-up of nitrogen molecules to form more complex materials, including 'tholins'. Reactions are symbolised by arrows. State-of-the-art photochemical models incorporate hundreds of reactions and dozens of compounds.

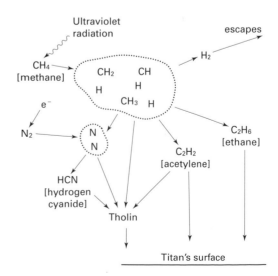

Another major product of the photochemistry is acetylene (also known as ethyne). According to predictions, there should be around one third as much acetylene as ethane. If it were produced continuously, it would form a solid layer on Titan a couple of hundred metres thick. Solid acetylene is explosively unstable. This fact was pointed out in the mid-1980s, prompting remarks that a space probe landing on Titan might detonate the whole moon! It was countered by the riposte that, if such were the case, some meteorite would have set the surface off long ago anyway.

The chemistry in Titan's atmosphere is difficult to reproduce exactly in a laboratory. It is not quite good enough to take a mix of nitrogen and methane and switch on a UV lamp. The catalytic effect of the walls of the reaction vessel is one problem. Another is that chemistry is affected by condensation. Some materials begin to condense out, depending on the temperatures and pressures in the vessel, whereas on the real Titan they may not. The tiny traces of oxygen that seep into the best-sealed vessel also affect the chemistry, although recent work in Paris is getting around this problem.

A tradition of chemical names

Acetylene is one of a number of chemical names, supposedly superseded, that have died hard. The International Union of Pure and Applied Chemistry designates this triple-bonded molecule 'ethyne' – the first of the alkynes, the next of which is propyne, followed by butyne. This systematic set of names is logical but for some reason, industrial chemists and atmospheric photochemists seem to prefer the old name. Similarly, 'ethylene' is the favoured, although formally incorrect, term for the double-bonded molecule more properly known as ethene.

Several of the compounds detected in Titan's atmosphere are well-known petrochemicals. Acetylene and ethylene are used on Earth in the manufacture of plastics. Propane and butane are fuels as, of course, is methane itself. Several of the hydrocarbons detected, such as diacetylene, are rather exotic and unstable in significant concentrations on Earth .

Titan's photochemistry is made rather more interesting by the presence of nitrogen. As well as mediating some reactions while remaining unchanged itself, nitrogen also enters the photochemistry directly. Although ultraviolet light does not split nitrogen molecules as easily as it breaks up methane, energetic particles lancing into

Table 3.1. *The composition of Titan's atmosphere.*
Many of the mixing ratios are strongly altitude dependent.
Representative values are given here.

Nitrogen	N_2	90–97%
Argon	Ar	0–6%
Methane	CH_4	2–5%
Hydrogen	H_2	0.2%
Ethane	C_2H_6	10 p.p.m.
Carbon monoxide	CO	~10 p.p.m.
Acetylene	C_2H_2	2 p.p.m.
Propane	C_3H_8	500 p.p.b.
Hydrogen cyanide	HCN	170 p.p.b.
Ethylene	C_2H_4	100 p.p.b.
Acetonitrile	CH_3CN	5 p.p.b.
Carbon dioxide	CO_2	10 p.p.b.
Cyanoacetylene	HC_3N	10 p.p.b.
Methylacetylene	CH_3C_2H	5 p.p.b.
Cyanogen	C_2N_2	5 p.p.b.
Water vapour	H_2O	8 p.p.b.
Diacetylene	C_4H_2	1 p.p.b.

Titan's atmosphere from Saturn's magnetosphere can do so. This mechanism produces excited nitrogen molecules with enhanced amounts of internal energy and molecules broken into individual nitrogen atoms or ions. These can combine with hydrocarbon fragments to make nitrogen-containing organic chemicals, known collectively as nitriles.

The simplest and most abundant of these is hydrogen cyanide – a molecule with one atom each of hydrogen, carbon and nitrogen. It is a potent toxin and has been used as such in gas chambers. Although hydrogen cyanide had been detected in deep space by its spectral signature, Titan's was the first 'planetary' atmosphere in which it was detected, when its presence was revealed by the infrared spectrometer on *Voyager 1*. More complex nitriles have been detected too, such as cyanogen and acetonitrile.

All this photochemistry is interesting for chemists but it only goes half way to addressing theories on the origins of life, a connection which is often used to justify the exploration of Titan. The problem is this. Barring some tiny traces we'll discuss later, Titan's atmosphere is bereft of oxygen or oxygen-containing compounds like water. This is a crucial point for two reasons. First, the chemistry that sustains life is mediated by liquid water; it requires liquid water as a solvent. Second, virtually every organic molecule of biochemical interest contains some oxygen – that means every amino acid, and therefore every protein and enzyme, every sugar, every fatty acid, and DNA itself. If there is no way to incorporate oxygen into the chemical chain in the atmosphere (and there isn't) then the organics that make Titan's atmosphere such a murky soup will only ever be sterile hydrocarbons and nitriles. However, all it might take to transform this stuff that falls manna-like from the sky is exposure to molten ice. Just add water, as if it were packet soup.

The outer layers of the young Titan would have been essentially an ammonia solution but in the present era Titan's surface is far too cold for liquid water. At 180 degrees below the freezing point, all water is rock solid. The technicalities that might allow liquid water to persist on Mars – thin films of water bound onto mineral grains and the antifreeze effects of salts – aren't enough to save Titan's water from the cold. Even the potent antifreeze ammonia doesn't open a loophole wide enough for liquid water. Thus, the big picture says that Titan has no liquid water and the chemistry stops when material from the atmosphere rains out onto the surface.

The same logic applied to Earth says that Earth is too cold for molten silicates – in other words, liquid rock. Happily, this is true by and large. But in particular regions at particular times conditions far out of the ordinary can take hold. Episodes of volcanism illustrate the point. Material erupted from Earth's interior as hot liquid can trigger chemical processes that do not usually occur. Most mineral deposits of economic value are formed by hydrothermal systems that sweat particular metals out of the rock and deposit them in concentrated form elsewhere.

An ammonia-rich ocean may still exist, deep in Titan's interior. A. Dominic Fortes at Imperial College London noted that conditions beneath Titan's surface may even today allow life to persist if it ever evolved when the ocean was exposed on the surface and able to interact with methane, organics, sunlight and incoming meteorites. Life down there would not be thrill-a-minute but the temperature, pressure and chemical environment would not be hostile to some of the more hardy and persistent life forms on Earth.

It is certain that comets and meteorites have slammed into Titan from time to time. When a very large meteorite or asteroid strikes a planetary body, most of its energy goes into a huge explosion, which excavates a crater and may throw material high into the atmosphere, or even into space. A proportion of the energy carried by the impactor goes into heating the target rock, some of which may melt. The larger the crater the greater the volume of rock that melts in the impact. One of the biggest impact structures on Earth (but sadly without an obvious crater shape) is the two-billion-year old Sudbury structure in Canada. The huge pool of molten rock lasted long enough for metal blobs in the rock to settle out at the bottom of the melt pool. Today these blobs, now long solidified, are one of the largest nickel deposits in the world.

Reid Thompson at Cornell University and his advisor Carl Sagan realised in 1991 that this kind of environment could exist on Titan. The ice would be melted in large impacts and would take thousands of years or longer to freeze again. So, locked up in Titan's icy surface, there is likely to be a plethora of molecules formed by the interactions of water with the ubiquitous hydrocarbons and nitriles. Amino acids are sure to be there but who knows how complicated the chemistry can get? A sophisticated surface-sampling mission to Titan in the future will be needed to find out.

Creating an atmosphere

Titan's atmosphere is big, not just in vertical extent but also in terms of mass. A big atmosphere prompts the big question – where did it come from? The question is particularly profound in the case of Titan. Not only is it the one satellite in the solar system with an atmosphere this big (its closest competitor, Triton, has an atmosphere less dense by a factor of about 100 000), it is also the possessor of the only other significant nitrogen atmosphere in the solar system besides our own.

To begin to tackle this enigma, we must first look at what astronomers generally believe to be the origin and evolution of planetary atmospheres in general. The solar system began life as the solar nebula, a large cloud of gas and dust, most of which ended up as the Sun. As this cloud contracted under its own gravity, it began to spin faster as a consequence of the conservation of angular momentum. The familiar analogy for this phenomenon is the way a spinning skater turns faster when she pulls in her arms. The fledgling solar system flattened into a disk, just like pizza dough. This version of events explains happily why all the planets go around the Sun in the same direction and more or less the same plane.

Some of the dust grains collided gently enough to stick together forming small clumps. The gravity of these clumps attracted more and more dust particles to assemble and coalesce into objects called planetesimals. As they grew larger, they became more efficient at collecting extra planetesimals and dust. Like competing species or corporations, a few of the fastest-growing clumps staked their turf, chasing away or absorbing their nearby competitors. Jupiter, the supreme fat cat of them all, even left a cloud of debris, the asteroid belt, as a warning to others not to intrude on its realm of the solar system.

But what was this 'dust'? Some remains to this day more or less unchanged since these earliest of times in the form of meteorites – some stony, some iron-rich, some with carbon-bearing material. This so-called refractory or involatile material can be likened to the 'Earth' element of the Ancient Greeks and was littered throughout the solar system. But 'Air' and 'Water' were pushed around by the fourth, 'Fire'. For the Sun had fired up at the centre of the cloud. It was sufficiently warm that water could exist only as a gas anywhere closer to the Sun than about 5 AU[1], a critical distance dubbed 'the snow line'. Within this boundary, water was much less abundant than beyond it. This may sound surprising, as our waterlogged world is only 1 AU from the Sun. But Earth apparently got its water at a later stage. The giant planets grew so massive they began to trap even that lightest of all elements, hydrogen – the main constituent of the Sun and of the nebula – as well as dust and ice. They created small sub-nebulae of their own, in which families of satellites formed. These satellites almost exclusively have ice on their surfaces – often earning them the label 'icy satellites'. The only prominent exception is volcanic Io.

Most likely, it was too warm for ammonia and methane to condense on the jovian satellites but, farther from the Sun, where Saturn and its moons were coming into being, ammonia as well as water might condense. Ammonia can be converted into nitrogen by sunlight or by the shock effects of lightning or meteorite impacts. So, if a lot of ammonia was acquired by Titan when it was in the process of forming, that ammonia could perhaps have been converted into the

---

[1] AU is the abbreviation for astronomical unit, which is Earth's average distance from the Sun: 149.6 million km (92.96 million miles)

nitrogen that exists today. The idea of enough nitrogen being caught directly in the ices that went into Titan is not plausible because it is such a volatile substance. Ammonia conversion seems to be the most obvious way for Titan to acquire a nitrogen atmosphere.

In 1997, there was a spurt of interest in the conversion of ammonia to nitrogen by the action of ultraviolet light when exobiologist Chris Chyba, then a professor at the University of Arizona, resurrected an idea of his former mentor, Carl Sagan. The problem they had puzzled over was that, with a faint early Sun, Earth should have been too cold for liquid water and for life. One explanation was that a much stronger greenhouse effect operated at that epoch but simply adding CO_2 didn't seem to work. Methane is a powerful greenhouse gas and would help but a combination of methane and ammonia would be better still. Even so, there were a couple of objections to this idea, which was originally suggested in the 1970s. The first hitch is that there are no significant ammonia-creating reactions on Earth, nor any geochemical evidence that ammonia was ever abundant. The second problem is that ammonia would be destroyed too quickly by sunlight.

Chyba and Sagan came up with a novel twist, namely that methane would be subject to photochemistry too, just as on present-day Titan, and would form a red haze. This haze layer would absorb blue and ultraviolet light high in the atmosphere (as it does on Titan) and so would protect the ammonia beneath. It was a neat idea but it did have a flaw. Chyba and Sagan didn't adequately account for the antigreen-house effect of the haze. Chris McKay, Jonathan Lunine and Lorenz showed that it would absorb and reflect a lot of the incoming sunlight, offsetting the greenhouse warming.

Though something of an aside, this story illustrates the kind of analogies between Titan and the early Earth that continually crop up and lend Titan its particular fascination.

The Bermuda Triangle and why Titan turned itself inside out

So far, so good for ammonia (and nitrogen) on Titan. But what of the methane? Methane is considerably more volatile than ammonia, so it doesn't automatically follow that Titan would acquire methane too. However, if water vapour is condensing into ice in the presence of methane, methane (and some other gases) can be trapped.

Water is strange stuff. Its polar molecules arrange themselves in a relatively open lattice when it freezes, which is why water ice is less dense than liquid water and floats. This lattice, under the right conditions, can trap within it an additional molecule, such as methane. The guest molecule is not chemically bound with the water, just physically trapped in a cage of water molecules – about six water molecules for every guest. This type of material is called a 'clathrate'. Methane clathrate can form on Earth deep beneath the ocean bed and sometimes appears in cores from deep-sea drilling. Small earthquakes and changes in temperature or sea level can cause the clathrate to decompose, releasing large amounts of methane, sometimes in catastrophically sudden events triggering undersea landslides that in turn inundate coastal settlements with tsunamis or 'tidal waves'. Clathrate decomposition has even been suggested as a cause for the various mysterious phenomena in the Bermuda Triangle – violent upwellings of bubble-laden sea might cause magnetic anomalies and ships would lose buoyancy in the froth, sinking instantly. Methane clathrates have also come to the attention of terrestrial climatologists in recent years since sudden releases of large amounts of methane, which is a strong greenhouse gas, could precipitate rapid climate warming.

Enclathration would be a way for Titan to have incorporated large amounts of methane when it formed, without the very low temperatures needed for methane to accumulate as a pure material. The enclathration idea is possibly supported by evidence that suggests a substantial loss of nitrogen from the atmosphere while methane was not lost in the same way (see later in this chapter). These disparate effects can be reconciled if most of the methane now present in Titan's atmosphere was locked in Titan's interior while the nitrogen was being lost. Much of the methane could have been locked up as clathrates in the interior for the first half billion years and only released after the strong solar wind that caused the nitrogen loss had died down.

This all sounds complicated but is actually straightforward. The satellites were originally assembled from smaller blocks or planetesimals. Like everything else with mass, these attracted each other, which is how the larger bodies formed. In the earliest stages, when proto-Titan was small, these planetesimals orbiting Saturn just blundered into one another and stuck. So the core of Titan started as a collection of these planetesimals – a mix of rock and ice clumped together. But as Titan grew, its gravitational pull increased and the material making its outer layers fell ever more energetically onto the surface. Although no-one knows what fraction of the energy is

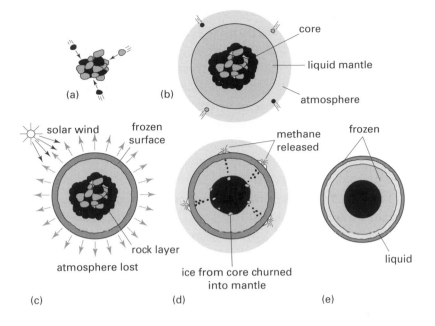

Figure 3.5. The formation of Titan and the origin of its interior structure. (a) Initially planetesimals made of rock and ice clump together slowly. (b) As the proto-Titan grows, incoming planetesimals release enough energy as they fall to melt ice so the accumulating body is liquid near the surface. It consists of a rock-rich mantle overlying a core of mixed rock and ice. The initial atmosphere of ammonia, methane, nitrogen and water vapour is formed at this stage. (c) The surface freezes over and the initial atmosphere is lost due to the action of an intense solar wind. Meanwhile, heating of the core due to radioactive decay softens the ice and makes the rock layer unstable. (d) The rock layer founders and the methane-bearing ice in the core is churned into the liquid mantle, releasing methane to the surface. (e) At the present time, the liquid mantle is mostly, but not completely, frozen.

retained by the growing satellite and how much is just blasted into space, eventually enough energy was liberated in the collisions to melt ice. When that happened, the rocky material would have sunk towards the centre. Thus, the new-born Titan would have had a core of rock and ice, overlain by a carapace of rock, surrounded by water – probably with an ammonia-rich atmosphere on top, slowly being converted into nitrogen.

However, after around 500 million years, Titan would have turned itself inside out. The dense rocky carapace would have foundered and collapsed into the rock–ice core, which had been warmed and softened by the decay of radioactive isotopes in its rocky component. It may be that this foundering, delayed by hundreds of millions of years, is what saved a large inventory of methane and delivered it to the surface after the nitrogen atmosphere had been largely lost.

As time went on and the body lost heat to space, the watery mantle would have frozen over. The peculiar properties of ice at the high pressures inside satellites, and with varying amounts of ammonia, mean that the last remaining liquid (probably there to the present day) has layers of ice both above and below it. Once the surface froze over, the atmosphere would have been largely isolated from the interior, apart from the odd squirt of volcanic gas.

Blast off!

Having got your atmosphere, the next trick is to keep it. Like many things, atmospheres keep best when cold. Sir James Jeans used his kinetic theory of gases as early as 1925 to prove that Titan was cold enough to hold on to an atmosphere for the age of the solar system, as long as the molecules in the atmosphere were not too light-weight. The sums don't work out for the warmer Galilean satellites of Jupiter, even if they managed somehow to acquire an atmosphere in the first place.

There is another factor at work that is likely to have acted in Titan's favour. NASA scientist, Kevin Zahnle, is a fan of giant impacts. He noted that the most dramatic ones, like those that formed the huge basins on the Moon, unleash millions of times more energy than the impact on Earth believed to have triggered the events that ultimately led to the demise of the dinosaurs. Impacts on this scale could blow off part of an atmosphere. Since a thinner atmosphere is easier to blow off than a thicker one, this process becomes progressively more effortless once it has started to happen. Every time an impact blasts away part of the atmosphere, it becomes possible for a smaller object to do the job as effectively on the next occasion. 'Atmospheric erosion' of this kind may be partly responsible for Mars's thin atmosphere.

Zahnle noted that comets would hit Titan rather more slowly than they would Jupiter's Galilean satellites because Saturn is farther from the Sun and comets gather speed as they approach the inner solar system. A slower impact is more likely to add material to an atmosphere than blow it away into space. Statistically, the mostly slow impacts on Titan would increase its atmospheric mass – plausibly to the mass we see today – while the more energetic impacts on the Galilean moons would destroy whatever atmosphere they had.

The larger Earth was better able to hold onto its atmosphere, although the energy of impacts would have been formidable. The last,

giant impacts on Earth around four billion years ago were energetic
enough to boil whatever oceans there were. It has been noticed that
the earliest known forms of life seem to be closely related to simple
present-day organisms that are very tolerant of heat and salt, often
found around volcanic vents. One idea is that these 'oldest ancestor'
organisms are not necessarily the 'first' organisms to have evolved but
are rather the 'sole survivors' of the last large impact that nearly ster-
ilised the planet. The first 500 million years of the solar system was a
dramatic time for Earth as well as Titan!

Goldilocks and the three phases

On our own planet, we are intuitively familiar with the way the atmosphere interacts with the surface. Our regime of temperature and pressure means that water commonly switches between its gas, liquid and
solid manifestations, and it's embroiled in two-way processes that
transfer it back and forth between surface and atmosphere. By contrast, the liquid form of rock – volcanic lava – is much scarcer and
localised. But translate to another world, where temperature and pressure are very different, and the processes that cycle substances
between surface and atmosphere operate in a very different and
apparently alien fashion. But however different other worlds may be
from our everyday experience, basic physics and chemistry gleaned
on Earth can tell us what to expect.

The typical pressure at the surface of Mars is only 7 mbar. Yet
water, even at only 0 °C, has a vapour pressure of 6 mbar. So on Mars,
the boiling point of water is perilously close to its freezing point. (The
freezing point stays essentially the same, except at the extremely high
pressures in the interiors of planets.) Liquid water at 'room temperature' (20 °C) on Mars would boil and freeze at the same time. It has to
boil since the vapour pressure is higher than ambient pressure. As it
boils, it takes away heat from the liquid, cooling it below the freezing
point. The ice thus formed would then sublimate away slowly.

Basically the same phenomenon occurs in space. Waste water from
the Space Shuttle, for example, is dumped overboard and in some
early flights this water dump happened when the pipes were cold.
Much of the water boiled, as intended, but some froze and was so cold
that it did not evaporate quickly enough, with the inconvenient result
that the on-board lavatory was blocked by ice and could not be used.
The solution was to turn the Shuttle so that the ice was exposed to the

Vapour pressure

Think about the surface of a solid or liquid substance. Molecules in the substance are in motion and those nearest the surface have a chance of escaping – flying free to create a gas (or 'vapour'). Equally, some molecules that have escaped will fall back onto the surface layer again. If the substance is in a closed box, a situation will soon develop when the rates at which molecules leave the surface and fall back are exactly balanced. In this state of equilibrium, the vapour will have a certain pressure, characteristic of whatever material it is. This pressure is called the saturation vapour pressure and its value varies according to temperature.

If the solid or liquid is not confined in a box, escaping molecules will gradually diffuse away. Molecules keep leaving the surface at a constant rate. This is how a liquid evaporates. If the material happens to be a solid rather than a liquid, the process is called 'sublimation'. The rate at which evaporation takes place depends on how big the difference is between the actual vapour pressure and the saturation vapour pressure. Imagine trying to dry washing on a very humid day. If the air is laden with moisture, it is very difficult to persuade the water in the washing to evaporate. Raising the temperature helps, and so does wind. Wind is effective because it blows gas molecules away, preventing vapour from building up locally.

Another way evaporation can be enhanced is to have the saturation vapour pressure exceed the total pressure of all gases present: under these circumstances the 'bubble' of vapour can just push the air away and a constant flow of vapour away from the surface can be maintained. There's a common name for this state of affairs. We call it boiling.

It is virtually impossible to keep a gas at a pressure higher than its saturation vapour pressure. If you try, it will tend to condense into a liquid or solid. For example, when the temperature of humid air drops, we get fog.

The saturation vapour pressures of most compounds change markedly with temperature. This is fortunate. It would be a considerable inconvenience if a cup of tea, made approximately at the boiling point of 100°C but drunk at, say, 50 °C, evaporated in a matter of minutes. The drop in temperature from 100 °C to 80 °C reduces the saturation vapour pressure by more than a factor of two, from 1000 mbar to 473 mbar.

Sun. As it warmed up, its vapour pressure – and the rate at which it evaporated – increased by a thousandfold or more.

Comets similarly have been preserved as ice blocks in a vacuum for billions of years because they are normally so far from the Sun. When comets venture into the inner solar system they get warmer and their ices (carbon monoxide, carbon dioxide and others as well as water) evaporate, forming a spectacular tail.

Neptune's satellite Triton formed relatively far from the Sun and acquired a substantial amount of nitrogen in its infancy – perhaps even as much as Titan. However, being several times farther from the Sun, Triton is also colder and modest heating by the Sun's weak rays is thwarted by Triton's bright surface, which reflects most of the sunlight back into space. Triton is therefore very cold (around 40 K). The saturation vapour pressure of nitrogen at this temperature is well under a millibar and indeed Triton's atmosphere was observed by *Voyager 2* to have a pressure of only a few tens of microbars. Most of Triton's nitrogen is condensed on the surface. Since 40 K is well below the freezing point of nitrogen, it takes the form of bright nitrogen frost.

If you moved Triton from Neptune (at 30 AU from the Sun) to Saturn (at 10 AU), in some respects it might turn into a very Titan-like object, since all the nitrogen frost would warm and evaporate to create a thick atmosphere. However, being only half as big as Titan, Triton would not be capable of holding on to its atmosphere as tightly and would lose it to space more quickly.

A similar chemical (rather than physical) temperature effect is seen on comparing the terrestrial planets Earth, Mars and Venus. All have about the same amount of carbon dioxide (CO_2) on their surfaces. On Earth most of it is thankfully locked up in carbonate rocks and CO_2 accounts for only a few hundred millionths of atmospheric pressure. On torrid Venus, the surface temperatures are high enough to decompose carbonates, pushing all the CO_2 into an atmosphere with a pressure 90 times Earth's. On Mars, at the other end of the scale, the temperatures are low enough for CO_2 to freeze out onto the surface and it forms seasonal polar caps.

Finding the conditions for liquid water on planets is often called the 'Goldilocks' problem – the trick is to be not too hot or too cold but just right. On Titan, at 94 K, the vapour pressure of water is immeasurably low and water ice is like rock on Earth. Carbon dioxide is similarly frozen rock hard. But methane is volatile enough to be a liquid and vapour on Titan though involatile enough (helped by enclathration, perhaps) to be found on Titan in abundance.

The notion that photochemistry involving the methane in Titan's atmosphere might produce liquids that could cover the surface was articulated as early as 1973. Hunten suggested that Titan might be covered by a kilometre of tar or asphalt. These initial scenarios were simply based on assessing the amount of ultraviolet light likely to have fallen on Titan over the age of the solar system. A more rigorous approach would need careful book-keeping on the variety of molecules at different levels in the atmosphere and the reactions between them. Darrell Strobel, now at Johns Hopkins University in Maryland, made one of the first attempts to construct a realistic simulation of Titan's atmospheric chemistry. He had been tackling the problem of photochemistry in giant planet atmospheres, and in Jupiter's in particular. He adapted his computer model of Jupiter to Titan. To his surprise, Strobel encountered enormous interest in his Titan work. A torrent of requests quickly stripped his supplies of printed copies of his paper. He attributes this great curiosity to the perceived link between Titan's organic chemistry and the chemistry leading to the origin of life on Earth, a connection most vigorously espoused by Carl Sagan.

Atmospheres in cyberspace

Photochemical models for Titan's atmosphere became more engaging after the *Voyager* encounters. In 1984, Caltech professor Yuk Yung developed a fairly elaborate representation. Models like this have an enormous number of inputs, such as the rates at which different chemical reactions proceed and the speeds with which their products diffuse through the atmosphere. Some of the reaction rates required as input can be measured in laboratories on Earth but they often have to be extrapolated to the very different temperature and pressure regime operating on Titan. These models also produce an enormous amount of data, including predictions for the concentration of each chemical at every level in the atmosphere.

While a lot of best-guess ignorance goes into creating such models, the data they produce can be tested. Yung's model was developed with the express purpose of 'predicting' the abundances of various molecules as measured by *Voyager 1*'s infrared spectrometer. It turned out that the model did quite a good job, suggesting that the chemistry it encapsulated was at least a fair description of what is actually happening on Titan.

Some compounds condense as liquids or solids and drizzle out of the atmosphere. If this didn't happen, they would build up to improba-

> **Photochemical models**
>
> The concept behind computer modelling is to create in cyberspace a realistic simulation of the processes nature is carrying out in the inaccessible reaches of interplanetary space. Since it is impossible to do the 'analogue' experiment of recreating Titan's atmosphere in a laboratory, researchers instead resort to digital techniques. They think of the atmosphere as a three-dimensional jigsaw of layers or boxes. All the properties of the molecules forming the atmosphere, the radiation impinging on them, and their mutual interactions, have to be embodied in tables of numbers and mathematical equations. Some of the numbers are reasonably well known. Others are guesses for unknown quantities. If you guess correctly, then the model should predict phenomena that can be corroborated against real observations. In theory, you can arrive at correct values for the unknown quantities by a process of elimination and refinement. In reality it is not so simple. It is much like the difference between the subtleties of a real live adventure and the limitations of a computer game, however clever the programmer.

ble concentrations. Naturally, the compounds in greatest production – ethane and acetylene – are among those that drizzle out the most. Yung noted that, over the age of the solar system, enough ethane ought to be produced to submerge the entire surface of Titan in a layer 600 m deep.

A Ph.D. student at Caltech at the time, Jonathan Lunine, realised that this quantity of ethane was significant and would form oceans, since ethane is a liquid at Titan's surface temperature. Better yet, ethane itself has a low vapour pressure and would dilute the methane, lowering its vapour pressure above the mixed ocean. A similar kind of effect can be observed in a domestic kitchen. Spill some syrup and note how much longer it takes to dry up than plain water. In technical terms, the effect can be put down to the lower vapour pressure of water over a concentrated sugar solution. Ethane would act on methane like sugar acts on water, preventing evaporation.

Voyager data showed that Titan's atmosphere was not saturated with methane vapour immediately above the surface. This had been one of the main objections to methane seas on Titan. But Lunine saw how everything might fit together. He envisaged huge oceans of methane slowly evaporating into the atmosphere, there to be converted irreversibly into ethane which drizzled back into the oceans.

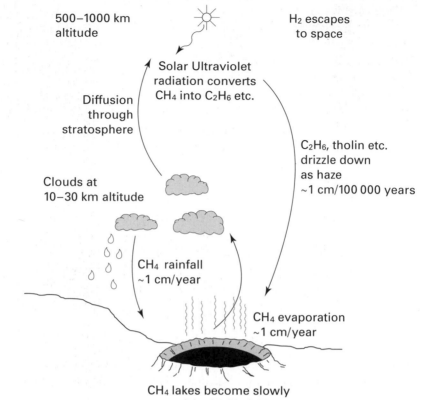

500–1000 km
altitude

H₂ escapes
to space

Solar Ultraviolet
radiation converts
CH₄ into C₂H₆ etc.

Diffusion
through
stratosphere

C₂H₆, tholin etc.
drizzle down
as haze
~1 cm/100 000 years

Clouds at
10–30 km altitude

CH₄ rainfall
~1 cm/year

CH₄ evaporation
~1 cm/year

CH₄ lakes become slowly
enriched with C₂H₆

Figure 3.6. The methane cycle on Titan, which is similar to the hydrological cycle on Earth. Methane circulates in Titan's lower atmosphere as a vapour, as cloud particles in liquid and solid form, and as rain or hail. A very small amount diffuses to high altitudes where it is destroyed by sunlight and converted into more complex organic compounds. These in turn drizzle irreversibly to the ground but in minute quantities compared with the amount of methane circulating between the atmosphere and the surface. Over time, the methane is depleted and converted into more complex molecules (principally ethane) on the surface. Hydrogen released in the photolysis of methane escapes into space.

Over time, the oceans became progressively richer in ethane and less volatile. The *Voyager* data weren't quite good enough to say exactly what the proportions of ethane and methane were but something between 25% and 60% ethane seemed about right. Given that the photochemistry model reckoned a production of about 500 m of ethane, the mixed oceans would be between 700 m (for the most ethane-rich case) and 4 km deep.

Lunine's short paper in *Science* magazine, 'Ethane ocean on Titan', written with his advisor David Stevenson and Yuk Yung, was one of the more influential and imaginative to be written about Titan. This

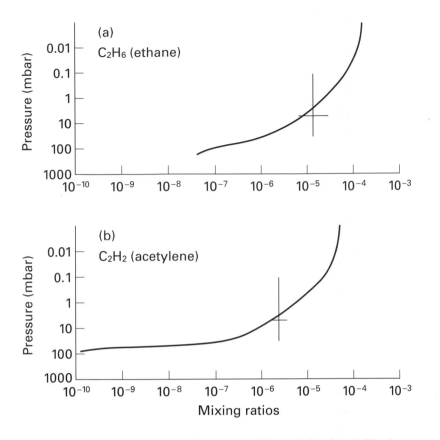

Figure 3.7. Matching calculations based on a model of the photochemistry in Titan's atmosphere with real data. The two large crosses represent actual measurements of C_2H_6 (ethane) and C_2H_2 (acetylene) with their range of uncertainty. The solid lines are calculations from a photochemical model developed by Luisa Lara and colleagues. The technique is to adjust the parameters used in the model until its predictions about how the concentrations of various compounds will vary with altitude match known data, such as the two points plotted here, for as many compounds as possible. Adapted from a figure by Lara *et al.* in *Planetary and Space Science*, vol. **42**, p. 5 (1993).

was how Titan could have seas to accommodate the methane expected to be there without contradicting *Voyager*'s observation. Of course, as we now know, this merely set up another conundrum for the early 1990s: how could such seas could be consistent with radar observations and with tides on Titan?

In the early 1990s, more models for the photochemistry were developed, one with many improvements over Yung's, by Frenchman Dominic Toublanc and another by Luisa Lara from the Instituto Astrofísica de Andalucia in Southern Spain, who was working for her doctoral thesis. Building a photochemical model is a great project for

a Ph.D. student. You have to find lots of factors to build into its compu-
tations and make the program work. You tweak the details until the
predictions it spews out fit the observations better than anything published before. When you defend your Ph.D. thesis, the committee asks
how your work could be improved. You outline why you were never
quite happy with, say, the way condensation is handled and explain
how for your postdoctoral research you would like to investigate this,
noting that better treatment of condensation will also be useful for
models of photochemistry on Earth/Mars/Jupiter etc.

This is all very well and, with enough good observational data for
comparison, much can be learned. But how close are these cyber cre-
ations to reality? How far can they be trusted, if many unknown parameters can be tuned so that the results fit only a few data points?
Changing any one of the dozens of reaction rates by a modest factor
can affect the results enormously and yet many of these reaction rates
are known with little certainty. Furthermore, in such a complex construction, it is an intractable job to test the sensitivity of the results to
each and every reaction rate. There is a sort of 'faith' that all the
errors and uncertainties will on balance cancel each other out and
that the result is a close proximity to the truth. There is no real way of
testing the results unless other scientists are given full access to the
computer code, which, since it embodies years of hard work, is rarely
granted. Unhappiness with this sort of inscrutable science is a major
theme of Clifford Stoll's book *Silicon Snake Oil* and we will come back
to it in the next chapter. (Stoll in fact worked with Tomasko and his
team in Arizona during the *Pioneer 11* mission.)

RALPH'S LOG. 1993.

*In 1993 I teamed up with Lara to write a paper documenting some of
the results of her model. I was hoping to find that it predicted less
ethane than Yung's model, which had given the only number in town
until that point. Whatever the ethane concentration in the ocean, the
ocean depth scaled directly with the amount of ethane. More ethane
meant deeper oceans. The radar data was difficult to reconcile with an
ocean 2 km deep on average, which would surely cover all of the
surface bar a few little islands. But perhaps if the ocean were only
300–500 m deep, the combination of deep crater basins and concealment of some of the liquid in a porous or cavernous material would
make the surface look 'solid' to radar. Lara's model indeed predicted
only 285 m of ethane. While the paper had the desired effect of*

chipping away at the bastion Yung had created of a global ocean, it was an excruciating experience. Day after day, as the manuscript was nearing completion, I would need to go through and change all the numbers by twenty or thirty per cent as Lara added yet another improvement to the model. When would it ever be right? This frustration was nothing to do with Lara – whom I was dating at the time. Tedious recalculation and an inscrutable sensitivity to the input parameters are necessary evils of this kind of science.

The crusade goes on. In order to tackle the complexity of Titan, new models are being built and their results explored, with different chemistries for different latitudes. As a counterpoint to the trend in complexity, Nick Smith, a British Ph.D. student studying in Paris with François Raulin in Paris, has used a simple model but has varied the uncertain reaction rates to show just how much 'slack' there is in these models.

A catastrophic climate history?

Titan may be very cold but it is as warm as it is partly because nitrogen and methane are highly effective as 'greenhouse' gases. Both are capable of condensing to form lakes and seas on the surface, or of dissolving in existing bodies of liquid. This means that the surface conditions can have a profound influence on the thickness of the atmosphere. This idea seems a little alien to us on Earth, until we remember that about half of Earth's greenhouse effect (without which conditions would be inhospitably cold) is due to water vapour. The amount of water vapour in the atmosphere depends, of course, on the surface temperature. On average, over the surface of Earth it amounts to one or two percent.

Mars is an even more pertinent example. Its atmosphere is in equilibrium with the surface. Even though the polar regions are in total darkness for months, the temperature stops falling when it gets to about 150 K because that is the temperature at which the atmosphere starts to freeze out onto the surface in the form of frost. In fact, during the course of the martian year, the atmospheric pressure changes by about 30% because much of the atmosphere freezes out to form seasonal frost caps in winter.

This coupling of surface to atmosphere raises some remarkable possibilities for Titan. A positive feedback could operate during climate fluctuations. Imagine that the Sun is a little fainter for some reason: the surface will naturally get a little colder. At lower temperatures, the vapour pressures of nitrogen and methane are lower too, so some of the atmosphere will rain out or dissolve in the seas on the surface and the atmosphere will get thinner. But a thinner atmosphere has a weaker greenhouse effect and so the surface will get a little cooler still. More nitrogen and methane will be lost from atmosphere to surface – and so on. Exactly what happens depends on the outcome of the competition between these two effects.

If Titan has relatively small reservoirs of methane and nitrogen on the surface in small and shallow lakes, there is not much capacity for dissolving or releasing gas. Under these circumstances, temperature changes on the surface provoke relatively minor changes in gas pressure. We could say that the system is well behaved. Crank down the Sun by 1% and it gets two degrees cooler while the pressure drops by a few hundredths of a bar. Crank the Sun back up again and conditions on Titan return to how they were before.

On the other hand, things get very interesting if Titan is endowed with vast supplies of volatile liquids. For a certain range of temperatures and pressures, its response to change is well-behaved. However, for some possible scenarios, the ocean/atmosphere system is unstable. Warm the surface a little bit and the pressure rises. The greenhouse effect intensifies and the surface warms more, releasing more gas, making it warmer still, until the oceans have all but boiled away. Rather than varying smoothly, the system jumps to a new stable configuration.

The Titan system also exhibits what is called hysteresis. Nothing to do with the hysterics induced in scientists by such bizarre behaviour, it's from the Greek words *hysteresis*, a shortcoming and *hysteros*, later. It means that changes are not altogether reversible. One can crank the Sun up by 1%, say, and conditions on Titan jump to a new state, tens of degrees higher and with a pressure of several bars – hardly in proportion to the change in circumstances. And yet, when the Sun's power diminishes again by 1%, the temperature drops by only a degree or two. But let the Sun dim some more and then the system jumps back to its old state. There are some values of the Sun's power output for which Titan has two stable states. Which one it happens to be in depends on what has happened previously. This is hysteresis in action. The effect is illustrated in Figure 3.8. The same kind of effect turns up in studies of Earth's climate. If Earth cools to the point at

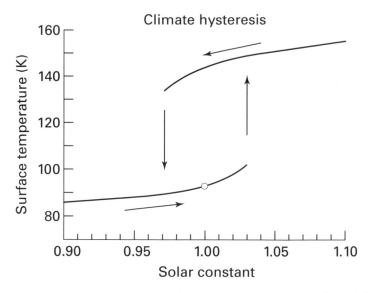

Figure 3.8. Hysteresis in Titan's climate. Starting with the lower curve towards the left of the diagram, imagine an increase in solar heating, which causes a rise in surface tempera-ture. The temperature rises gradually at first, following the lower curve, but then jumps to the higher curve when the runaway greenhouse effect takes over. A subsequent reduction in heating does not result in the climate immediately returning to its former state. States between the two curves can never be attained. The circle indicates the present state of Titan. The curves are calculated assuming that large quantities of potential greenhouse gases, such as methane and nitrogen, are condensed on Titan's surface. An increase in heatflow from Titan's interior, or of the abundance of greenhouse gases, could have a similar effect to an increase in solar heating.

which the polar ice sheets extend towards the equator as far as lati-tudes of about 45°, the vast expanse of whiteness reflects the Sun's warming rays and Earth cools even more. Eventually, it becomes covered entirely with ice and is difficult to thaw out again.

It needn't be the Sun that is responsible for such effects. Any other warming influence will do, such as geothermal heating. The presence of an additional noncondensible greenhouse gas like hydrogen might have a similar effect. However, the Sun is an important factor because we know it is increasing in luminosity by about 10% every billion years. Sooner or later, if Titan is volatile-rich, it may get pushed over the edge towards a catastrophic runaway of the greenhouse effect. It could even be that the 1.5 bar pressure we see today is Titan on the brink of such a dramatic conversion. In a century the pressure could be 4 bars. By the same token, a volatile-rich Titan probably had a rather thinner atmosphere in the past, since a large methane ocean would be better able to dissolve nitrogen out of the atmosphere. It is possible that Titan's climate history has been catastrophic, oscillating between thick-atmosphere episodes and thin-atmosphere episodes. The *Cassini* mission may collect enough clues from Titan's surface uncover the truth. In Chapter 5 we'll look at how that might be done.

Add a little carbon dioxide

So far, we've discussed the bulk components of the atmosphere. But even the gases present in tiny amounts can have an important story to tell. The saturation vapour pressures of water and carbon dioxide are tiny at Titan's surface. They would be even lower up in the atmosphere near the tropopause at 70 K. Things improve somewhat in the stratosphere, hundreds of kilometres above the surface, where the temperature reaches 180 K. However, there is no way for material to get there from the surface. Even if saturated vapour at 94 K could be transported upwards from the surface, the CO_2 and H_2O would be frozen out of the atmosphere at the tropopause, which acts as a 'cold trap'. The same thing happens on Earth. The Earth's stratosphere is quite dry, because water vapour from the moist surface is caught at the tropospheric cold trap.

All that being so, why is there any CO_2 at all in Titan's atmosphere? Any put there by some mechanism would be removed in the form of snow or frost as circulation brought it to the cold trap. Either CO_2 is being continuously added to Titan's atmosphere from outside, or it has to be formed in the atmosphere by some chemical process. Is continuous creation of CO_2 in the atmosphere a plausible theory? Only one way to make CO_2 photochemically is known. It involves a reaction between carbon monoxide (CO) and OH radicals. CO is more volatile than CO_2 and doesn't condense out at the cold trap in the same way. There is plenty of CO. But OH could only come from the action of sunlight on water vapour. And that's where the idea comes unstuck. Water vapour suffers from the same problem as CO_2 and would get snowed out at the cold trap. An inevitable conclusion follows this piece of reasoning: either CO_2 or water (or both) have to be added to the atmosphere *from outside.*

Bob Samuelson worked out that this was possible and made a first attempt to calculate how much water was needed. He guessed that it was being supplied by icy meteors 'boiling away' as they entered the upper atmosphere. We say 'boiling' because they wouldn't usually get hot enough to 'burn up' the way meteors do in Earth's atmosphere. Meteoric chemistry sounds rather far-fetched, and indeed it is, but here was an imaginative explanation. The only real test was whether it worked.

Sodium, the metallic element in common salt, is highly reactive. It is fairly soft and melts at only a few hundred degrees. However, its vapour pressure is still tiny at Earth's surface temperatures. You might therefore think it somewhat surprising that wispy layers of sodium vapour can be detected at altitudes of nearly 100 km in Earth's atmosphere. Even iron has been detected at such heights. This is the level where most meteors burn up. Some of the rock and metal vapours condense to form a little smoke trail but some sodium and iron atoms left floating around can be detected from the ground. There is probably no ice since it is warm enough at Earth's distance from the Sun for a meteor to lose any ices it may have had directly into the vacuum of space long before it collides with Earth.

It requires powerful lasers beamed up from the ground to make the atoms of sodium and iron fluoresce. They emit light of particular recognisable wavelengths, the familiar bright yellow of street lights in the case of sodium. This light can be detected on the ground with a telescope equipped with a narrow-band filter. The experimental set-up is called a lidar, an acronym derived from LIght Detection And Ranging , inspired by RADAR for the RAdio equivalent. And as with radar, by timing the interval that elapses from blasting the laser pulse into the sky until the faint fluorescent echo is received, atmospheric scientists can determine how far away the sodium layer is.

RALPH'S LOG. 1993.

I got the opportunity to see one of these atmospheric lidars in operation during a field trip at the first European Research Course in Atmospheres, run from the University of Grenoble in southern France. We visited the Observatoire de Haute-Provence. It was observations made here that led to the first report of an extrasolar planet being detected only a couple of years later. It was quite a magical and surreal experience to gaze at the dark night sky, replete with a multi-tude of stars (notable because this region has rather less light pollution than most Europeans are used to), the air scented with lavender and thyme, and to see a brilliant green shaft of light lance up from an observatory dome.

Viewed in this fluorescent light, the notion of meteoric chemistry on Titan no longer seems quite as bizarre. One outstanding puzzle was, 'Where is the water?'. To work out how much water should persist in Titan's atmosphere given the known amount of CO_2 required the application of a detailed photochemical model. Not only that, but it would be important to know the altitudes at which the water was deposited. That required a model of the sizes and velocities of meteors and how they burned up in the atmosphere.

Coupling together all these effects would be a big project. The first attack on the problem was a joint effort. Lorenz teamed up with Mark English, another Ph.D. student at the University of Kent in Canterbury and postdoctoral researcher Paul Ratcliff. They worked out the likely meteoroid population, their speed distribution, and the burn-up profiles in the atmosphere. The profile of deposition in the atmosphere was then passed to Luisa Lara, who used her photochemical model to work out the resulting gas abundances.

Without any meteoric input, the CO_2 abundance would be limited to about 1 part per trillion (10^{-12}) because of condensation at the cold trap, far smaller than the part per billion or so determined from the *Voyager* and ground-based observations. Billion or trillion, it's still a trace gas, like CFCs in Earth's atmosphere. It's remarkable that we can study the chemistry of such tenuous trace gases on a world a billion miles away.

If the meteoroids were made entirely of water, it would need many (tens to two hundred) times the expected flux of dust-sized meteoroids to provide the water needed to make the observed amount of CO_2. That there should be so many seemed a little improbable, since measurements made by *Pioneer 11* showed that the amount of interplanetary dust was fairly constant out to Saturn. There was also another problem. Even though much of it would be used up in making CO_2, and some would condense, there would also be a significant amount (tens or hundreds of parts per billion) of water vapour in the stratosphere. Water vapour hadn't been detected in the *Voyager* infrared data, though that didn't necessarily mean there couldn't be at least a little. But how little? Athena Coustenis, who knew the *Voyager* data about as well as anyone, suggested one or two parts per billion as an appropriate upper limit.

This was all something of a puzzle. However, the Canterbury/ Andalucia team came up with a suggestion. Streams of meteoroids the size of dust grains are left behind in the wake of comets. But comets aren't made of pure water ice. In fact the images of Halley's Comet from the European cometary probe Giotto confirmed that they are positively filthy and data analysis showed that CO, CO_2, HCN and other gases were being released. If the meteoroids at Titan were made from anything like cometary material, then about 3% of the incoming mass would be CO_2. It would take only one tenth as many meteoroids with this composition to account for the CO_2 in Titan's atmosphere compared with the numbers needed if they were solid water. At the same time, the water abundance in the atmosphere would be ten times less, in better agreement with the lack of any detection in the *Voyager* data.

Even allowing for some latitude in the photochemical model, this theory still called for a flux of dust particles several times greater than the quantity expected. One possibility was that fine ejecta created by impacts on Saturn's outer satellites Hyperion, Iapetus and Phoebe might enhance the flux of meteoroids. At first sight, the rings might appear to be a more obvious source but, while it is easy for dust from, say, Hyperion to lose energy and spiral in towards Titan, it is more difficult to push dust outwards from the centre of the saturnian system against the gravity of a giant planet.

Ultimately, the water vapour in Titan's atmosphere was detected. This took some unprecedented observations and a Herculean analysis effort led by Coustenis. The observations were made with ISO, the European-built Infrared Space Observatory. While the telescope on this satellite, launched in 1995, had a mirror only 60 cm across (a quarter the size of Hubble's), it was chilled to an ultracold 4 K with liquid helium to make it extremely sensitive to faint infrared emissions. The infrared spectra it took were of better signal-to-noise, and far higher spectral resolution, than those made by *Voyager*, even though ISO was millions of times further away from its target.

By adding together several hours of observations from ISO taken in December 1997, shortly before ISO ran out of its helium coolant, Coustenis was able to find two spectral lines of water. Taken together, these suggested that the water abundance was about 8 p.p.b. at an altitude of 400 km, roughly in line with the expectations of the cometary dust theory. The new observations prompted a flurry of theoretical work, taking the same approach as the Canterbury/Andalucia team had done several years before and leading to essentially the same conclusions. The combination of *Cassini*'s sensitive infrared spectrometer (CIRS) and the *Cassini* Dust Analyser (CDA), which can measure the composition of dust particles, as well as their size and velocity, should nail the problem once and for all.

Eyeing up the isotopes

In the 1990s astronomers began to exploit parts of the electromagnetic spectrum that had barely been touched before. One of them was the millimetre-wave region. It falls uncomfortably between the far-infrared and radio. Far-infrared light is handled by techniques familiar to optical astronomers involving shiny mirrors whereas radio waves are more the purview of electrical engineers. Millimetre-wave telescopes are big dishes, like radio telescopes but they have to be much smoother and shinier in order to focus the radiation properly. The roughness permissible scales with the wavelength being observed.

Around 1998, millimetre-wave astronomy delivered an astonishing result concerning the isotopes of nitrogen in Titan's hydrogen cyanide, HCN. On Earth, every 278th nitrogen atom has an extra neutron, making its atomic weight 15 instead of 14. If a hydrogen cyanide molecule contains one of these heavy atoms of nitrogen instead of an ordinary one, the molecule vibrates at a slightly different frequency. We can imagine the molecule as a set of three ball bearings of different masses, which represent the hydrogen, carbon and nitrogen atoms, connected with springs, which represent the chemical bonds between them. The chemical bonds, mediated by the clouds of electrons that surround the nuclei of atoms, are the same regardless of whether the molecule contains an ordinary nitrogen atom or a heavy one. However, if the nitrogen ball bearing is one of the slightly heavier type, it will wobble a little slower than its N^{14} fellows. This difference in frequency can be detected in the millimetre-wave spectrum.

Amazingly, it turned out that Titan has 4.5 times more N^{15} atoms relative to N^{14} than Earth. Since all the planets were made from the same well-mixed cloud of gas that formed the Sun, there is no obvious way to add extra N^{15}. The only reasonable explanation for the N^{15} enhancement is that Titan started out with lots and lots of nitrogen, including both N^{14} and N^{15} in their primordial ratio, and much of the N^{14} has been lost. To arrive at the present proportions, Titan must have disposed of about 30 atmospheres-worth of molecular nitrogen (N_2), though with the margin of error on the approximate calculation, it may be as few as five or as many as 700. Some of the processes by which atmospheric gases can be lost to space depend on the mass of the molecules but this is a staggering amount to shed and the solar system environment we see today cannot account for it. The young Austrian scientists Helmut Lammer and Willi Stumptner at the Austrian Academy of Sciences in Graz looked at every possible process and found no credible candidate, unless the Sun's early episode of strong solar wind had been much longer and stronger than had been previously thought.

However, there is a twist. Carbon and hydrogen also have stable heavy isotopes. C^{13} in Titan's methane is not nearly as enriched as N^{15}, which suggests that methane was somehow protected from the early winnowing. This in turn means that the methane must have been released later from its icy cage. So far, the only explanation that has been advanced is the core inversion idea discussed earlier in this chapter. Though uncommon, C^{13} and N^{15} are stable. There also exist less stable isotopes that decay radioactively into another element over time. How fast this happens is described by the 'half life', the period over which half of the atoms will have metamorphosed. The plutonium-238 that powers *Cassini* has a half-life of 87 years; each atom of Pu^{238} decays into an atom of uranium. Hydrogen has an unstable isotope called tritium (H^3), which has a half-life of only 12.3 years. Since tritium is used in the triggers of hydrogen bombs, this short half-life means that these triggers must be replaced regularly. Carbon has an unstable isotope, C^{14}, with a half-life of 5570 years.

Of course, since their half-lives are short compared with the age of the solar system, these isotopes should not exist in nature and generally they don't. Pu^{238} and tritium have to be manufactured in nuclear reactors. C^{14} is one of the exceptions to the rule. It is produced in Earth's upper atmosphere by the interaction of energetic cosmic rays with nitrogen atoms. Since it is being produced at a more-or-less constant rate, and decaying away at an equal rate, its abundance in Earth's atmosphere hardly changes. Living things exchange carbon atoms with the atmosphere. This happens mainly because plants take up CO_2 and animals eat plants. So when plants and animals are alive they have the same proportions of different carbon isotopes. However, when they die and stop exchanging carbon with the atmosphere, the C^{14} atoms keep decaying but are not replaced. From then on, the C^{14}/C^{12} ratio drops, at a known rate. The present value of the ratio in a sample of organic material tells us how long has passed since it was alive. This is the principle behind the technique of radio-carbon dating. Very few compounds have spectroscopic signatures that allow useful isotopic measurements to be made from a distance. However, if a sample of the material can be obtained and inserted into an instrument called a mass spectrometer, the relative numbers of atoms with different atomic weights can be measured in a laboratory.

The noble gas argon has several isotopes and will be a useful tracer of atmospheric evolution on Titan. Two isotopes, Ar^{36} and Ar^{38}, occur naturally. Chemically, argon hardly reacts at all, making it a useful tracer. Physically, it acts a little like molecular nitrogen and should have been trapped in ice with a similar efficiency. If Titan's atmosphere came from nitrogen trapped in ice, there should also be a large amount of argon delivered in the same way. If so, the proportion of argon in the atmosphere today should be several per cent. The Ar^{36} /Ar^{38} ratio, if it can be measured by the *Huygens* probe, may give a little more information on the rate and timing of the nitrogen loss process.

Argon has another isotope, Ar^{40}. This is generated from the decay of potassium-40, with a half-life of 1.3 billion years . The solar system is about three of these half-lives old, so there is only about 12% left of the initial amount of K^{40}. Chemically, potassium is an alkali metal. Like sodium, it is found in rocks and in soluble salts. All the potassium in Titan will be present either as rock in Titan's interior, or as dissolved salts in Titan's ice crust. Either way, any argon released by potassium decay after Titan's surface froze solid should be trapped in the interior.

If we detect Ar^{40} in Titan's atmosphere, this means that the argon released in Titan's interior found a way to get out, somehow being vented through cracks or spewed out in volcanoes. This brings us back to methane.

One consistent scenario to explain the isotope ratios in Titan's atmosphere is that the nitrogen atmosphere was formed quickly and then part of it was lost quickly. In this picture, the methane was added to the atmosphere later on, perhaps as Titan's core overturned about 500 million years after it formed: if the methane were delivered too early, it would have suffered from the same loss process that changed the proportion of nitrogen isotopes. If the methane came late, or is even being delivered to the surface now, then Ar^{40} should be present in the atmosphere.

So much methane, so little carbon monoxide

Carbon monoxide is an odourless, colourless gas, made from one atom each of carbon and oxygen. It has the unfortunate property that it binds very tightly to haemoglobin. This red molecule is what makes blood red and in red blood cells it transports oxygen to muscles. Carbon monoxide, because it binds so strongly, prevents haemoglobin from carrying oxygen and it is therefore a deadly poison.

Carbon monoxide was first detected in Titan's atmosphere in 1982 by Barry Lutz of Northern Arizona University in near-infrared spectra taken from the ground. The 2-micron spectrum suggested an abundance of 50 p.p.m. in the lower part of the atmosphere. In 1987 French millimetre-wave measurements complicated things by finding a factor of ten less. These measurements were more sensitive to higher levels in the atmosphere and some theorists came up with dubious schemes to explain the discrepancy, suggesting that CO was washed down by raindrops, depleting its concentration in the upper atmosphere. However, another set of investigators, Ph.D. student Don Gurwell and his advisor Duane Muhleman of Caltech, made new millimetre-wave measurements with an interferometer and got a stratospheric abundance of CO similar to the level determined by Lutz, suggesting the same amount of CO throughout the atmosphere after all. But that amount is small.

In contrast with Titan, comets are rich in carbon monoxide and carbon dioxide but have very little methane. While a 'late veneer' of comets has been postulated as the origin of Earth's oceans and early atmosphere, the notion doesn't seem so plausible as an origin for Titan's atmosphere. If comets were the source, what happened to the CO? Was a substantial amount lost along with much of the original nitrogen?

Alternatively, perhaps CO has never been there in great quantity. The chemistry of carbon, hydrogen and oxygen in space distributes the atoms of these elements between various compounds in a way that depends on the pressure and temperature conditions, and the relative amounts of these and other elements. With enough hydrogen and with high pressure, as perhaps would have been the situation in the centres of the protoplanetary clouds from which the giant planets formed, carbon will take up as much hydrogen as it can to form methane (CH_4) and oxygen will absorb a full quota of hydrogen to form water (OH_2 – better known as H_2O) leaving none to mate with carbon to form CO.

The remaining hydrogen will be left as molecular hydrogen, H_2. The giant planets seem to fit this picture. So has Titan never had much CO because it formed like a little giant planet, minus the hydrogen its weaker gravity could not retain?

For the time being, the source of Titan's carbon monoxide remains an unsolved enigma.

The road to El Dorado

We are still grappling to understand Titan's atmosphere. The question of its origin has not been resolved unequivocally, although the weight of opinion is leaning on the ammonia conversion possibility. The bulk composition of the atmosphere is not quite pinned down. The argon abundance in particular is a gap in our knowledge. Measuring this with *Cassini* should answer the origin question. The cataclysmic story behind the nitrogen isotopes is still quite a new mystery; measurement of noble gases by the *Huygens* probe will help establish what happened. Our gradual progress towards the truth is a familiar route trodden regularly by astronomers. Every new and unexpected observation leads to a rash of exotic ideas to explain it. Then more observations make it clear which were the blind alleys and which is the real road ahead.

Murky meteorology

To judge by his writings, Christiaan Huygens, discoverer of Titan, would not have been surprised to learn that his world is both Earth-like and alien where weather is concerned, with wind and clouds, possibly even rain and oceans. Prophetically he speculated as follows.

> But since 'tis certain that the Earth and Jupiter have their Water and Clouds, there is no reason why the other Planets should be without them. I can't say they are exactly of the same nature with our Water; but that they should be liquid their use requires, as their beauty does that they be clear. For this Water of ours, in Jupiter or Saturn, would be frozen up instantly by reason of the vast distance of the Sun. Every Planet therefore must have its waters of such a temper, as to be proportion'd to its heat: Jupiter's and Saturn's must be of such a nature as not to be liable to Frost.

Not on Saturn itself as Huygens imagined, but remarkably close on its largest satellite, there is indeed a fluid substance not liable to frost that plays the role of water on Earth – methane. But neither Huygens nor his twentieth century successors anticipated the obscuring haze that dominates Titan's weather without exception, overshadowing the landscape with a murky veil. On the one hand, its impermeability to visible light is an irritating inconvenience for astronomers. On the other, it makes Titan all the more complex and exciting.

A hint of clouds

> Thy cheeks look red as Titan's face, Blushing to be encounter'd with a cloud
>
> W. Shakespeare – *Titus Andronicus*

The first inklings of hitherto undetected phenomena are often vague, ambiguous and generally disbelieved. Bernard Lyot and his sharp-eyed colleagues professed to seeing variable markings on Titan's disk during the 1940s and, earlier on, Comas Solà's 1907 drawing had Titan with two brightish spots. We may never know whether these observers recorded something real, but it is not unnatural to expect clouds in an atmosphere. When Dale Cruikshank and Jeffrey Morgan reported in 1980 that Titan's brightness probably varied, they suggested clouds as a possible explanation. But is wasn't clouds on this occasion. It later turned out that their signal was probably dominated by radiation from the surface.

The views recorded by the *Voyager* cameras were particularly disheartening for the seekers of clouds. None were obviously visible. Re-analysing the *Voyager* images some years later, Dan Wenkert and Glen Garneau of JPL made a conference presentation claiming that they had detected faint clouds on Titan, moving at speeds of 25–40 m/s, in a prograde direction (i.e. in the same direction as the rotation of Titan itself). However, this analysis relied on dramatically enhancing the contrast of the images. The alleged cloud features could simply have been noise, if viewed in an uncharitable light.

There was one other *Voyager* observation that could possibly be a hint of clouds. It was a subtle effect in the radio occultation experiment. The general trend in the signal indicated, as expected, that the atmosphere gets denser the deeper you descend into it. But superimposed on the general trend were some oscillations, called 'scintillations'. These were like the twinkling of a star seen through Earth's atmosphere. One possibility was that clouds had interfered with the

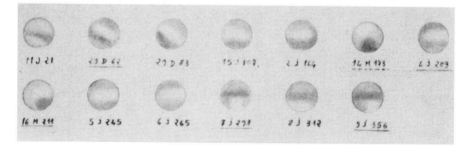

Figure 4.1. Drawings of Titan made by Bernard Lyot and colleagues using the 600-mm (24-inch) refractor at the Pic du Midi Observatory in France in the 1940s. Drawings were made independently by five observers and compared afterwards. These final drawings only show what had been detected by several observers. They are arranged in order of Titan's position in its orbit.

signal. Another was that small fluctuations in air temperature, propagating upwards as so-called 'gravity-acoustic' waves, were responsible. Of the few scientists who examined the data the majority rejected clouds and favoured the second interpretation. The first Earth-based images that might have revealed details on Titan's surface (see Chapter 2) showed features that could have been clouds, but there was no way of discriminating between clouds and surface features with single images.

One of the big hopes with the 1994 Hubble Space Telescope observations was that they would be able to detect clouds moving. But there was to be no such luck. Smith and colleagues stared at the images for hours. Certainly there were spots dancing around but they could have been noise. There was nothing obvious moving from frame to frame and different people saw different things.

RALPH'S LOG. MAY 1995.

After we had perfected our map-making programs and written up the results as a paper, ultimately published in 1996 in the journal Icarus, I set to work hunting for clouds in our HST data. Now that we had a map of Titan's surface, I tried subtracting the 'known' bright surface features from the individual images (which had already had the average 'haze' subtracted from them) to see what was left. Of course, after subtracting the haze (85–90% of the light) and the surface features (8–10%) only a few per cent of the light was left. Would it be anything but noise?

As I stared at the images, using different colour scales, watching the frames as movies, drawing graphs of cuts across lines of latitudes, it seemed that something stood out, moving west to east at the north-eastern edge of the bright continent. The cloud I thought I saw was a barely detectable signal in three of the frames – perhaps a 95% certainty that something was there. But in one frame there was undoubtedly something – a solid detection, if cloud it was. Could it have been a cosmic ray hit on the CCD camera? These are easy to spot against the blank background of space but harder to identify if superimposed on the image of a bright extended object. However, the feature also appeared in the same place in a separate red-filtered image. A cosmic ray hit would only affect a single frame and the red image at 673 nm barely sensed the surface. I was convinced, but could I be persuasive?

I presented these results as a poster at the 1995 meeting of the American Astronomical Society's Division of Planetary Science, which happened to be in Hawaii. (I recall that an extra effort of will was

*required to stay in the lecture room when the beach, coral reef and sea
turtles beckoned 200 yards away.) I didn't submit my findings for
publication because we were hoping to get ironclad evidence of clouds
the following year. A few months earlier we had submitted a proposal
to the Space Telescope for 20 looks at Titan. With that amount of data,
and our experience with the 1994 observations, we'd be sure to get
better evidence, we thought.*

*Unfortunately, our proposal was not accepted. The feedback from
the telescope time allocation committee declared that the probability of
detecting clouds was small because there was no evidence for them,
so precious Hubble time should not be wasted on a fruitless wild goose
chase. Ironically, a mere 30 miles away from my poster, and only two
weeks before, the light bearing conclusive evidence for clouds on Titan
bounced off the 3.8-metre mirror of the United Kingdom Infrared
Telescope (UKIRT) and into its powerful infrared spectrometer. The
data were written to tape but then lay in a desk drawer for the next
two years.*

Captured – a cloudy day

The arrival of conclusive evidence for clouds on Titan in many ways
mirrored the discovery of the atmosphere. Images of Titan's disk,
pushing the limits of resolution and sensitivity, gave tantalizing hints
that were persuasive to the observers themselves. It would be spectroscopy that provided the unambiguous and irrefutable evidence for
clouds.

Caitlin Griffith and Toby Owen had observed the radiation from
Titan in the part of the spectrum around 3 microns. It had not been
studied much previously because of Titan's extreme faintness at this
wavelength. But there was a methane 'window' here and it might be
possible to sense the surface. There is precious little sunlight at 3
microns, so observations demand a large telescope and sensitive
instrumentation. The spectrometer on UKIRT, kept extra cold to
reduce noise, could handle the job.

In contrast to our experience with visible light, ice is very dark at
these wavelengths, while rock is bright. With this knowledge, Griffith
and Owen hoped to discover something about the composition of
Titan's surface. When they looked at the data, it didn't make sense. On
one night, there was more light at 3 microns than could possibly come
from the surface, even if it were perfectly white. Titan was also
brighter at 2 microns than it ought to be. There was by now a well-

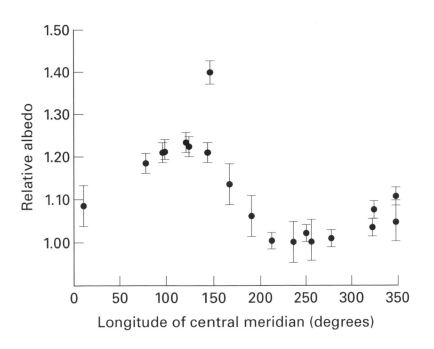

Figure 4.2. The variation in Titan's brightness as it rotates, measured at a wavelength of 2 microns. This light curve was assembled by Caitlan Griffith and colleagues from their own observations, supplemented by data from Mark Lemmon and Athena Coustenis. The rounded peak in the curve at about 110° longitude corresponds with the well-known bright region on Titan. The anomalous data point at longitude 140°, taken on the 4th of September 1995, is due to the exceptional presence of a bright cloud. Adapted from a diagram by Griffith *et al.* in *Nature*, vol. **395**, p. 575 (1998).

established light-curve at this wavelength, with variations in brightness repeating regularly and amounting to about 30%, but the data point for the 4th of September 1995, obvious in Figure 4.2, was about 100% brighter than expected, even though it wasn't taken at the brightest point of the regular light-curve. A large cloud system could be responsible.

The proof was in the subtle shape of the spectrum. At some wavelengths methane is strongly absorbing and we see only the light scattered from haze above most of the atmosphere. At others, methane absorbs only weakly and some of the light we see has come through from the surface. However, there is no sharp boundary between these regions of the spectrum. The degree of methane absorption changes steadily with wavelength from the centre of a 'window' to the centre of an absorption band. In these changeover wavelengths, we are seeing infrared light that gets through most of the atmosphere, but not quite all. Crucially, when the extra brightness in the 2 and 3 micron windows was recorded, the spectrum in the half-way regions was brighter too.

The information contained in the spectrum allowed Griffith to place the cloud quite firmly at an altitude of 10–15 km. The cloud had to cover about 10% of Titan's disk. Such dramatic clouds seem to be rare. By the year 2000, there had been several good looks at Titan, by Roland Meier (and Toby Owen), by the Arizona group, by Coustenis and colleagues in Paris, and by the Lawrence Livermore group using the giant Keck telescope in Hawaii. None saw anything like the huge 1995 cloud coverage.

In 2000, Griffith and colleagues reported the results of a more detailed analysis of their spectral data. This time they pushed their data to the limit, arguing that, in addition to the notable 10% coverage 'event' of 1995, there were regular 1% cloud features that persisted for only a few hours at most. Short-lived clouds were consistent with the idea that Titan's atmosphere might be supersaturated with methane; cloud droplets would grow quickly and fall as rain. It was possible after all that the 'noise' in the earlier HST images had really been short-lived clouds.

Titan weather forecast – mostly dry

There is roughly a metre of rainfall every year in the USA as a whole. In the wettest parts of the world, like the lush west coast of New Zealand, perhaps 6 m of rain fall annually. But let's take a metre as an average. One metre corresponds to 1000 kg of water on every square metre of surface. Since the oceans are not drying up, rainfall must balance the amount of moisture they lose by evaporation, taking a global average. Energy is needed to make water evaporate and that energy comes ultimately from the Sun. To evaporate 1000 kg of water over a year requires about 70 W/m^2. This latent heat is removed from the surface by the evaporation process and is released higher in the atmosphere as the liquid condenses to form clouds and rain. The heat removed from the surface must of course be less than the sunlight absorbed by the surface, which is around 260 W/m^2. Ignoring seasonal variation, Earth as a whole is not warming or cooling significantly so the 260 W/m^2 must be balanced by losses. Roughly a quarter of the absorbed sunlight goes into evaporation while the rest is either transported upwards by the convective motion of warm dry air or is radiated directly into space. Ultimately, the heat transported upwards as water vapour and warm dry air is radiated to space as well so that Earth and its atmosphere remain in thermal balance.

The same kind of calculation can be done for Titan. On Titan only around 0.4 W/m² of solar energy reaches the surface. Even if all of that heat went into evaporating methane, which has a latent heat smaller by a factor of five than water and a density half that of water, it is only enough to evaporate a 5-cm layer of methane, which sets an upper limit on the amount that can fall again as rain in each (Earth) year. Since some heat must be radiated and convected away, 5 cm is a wild overestimate and 1 cm seems a more likely value for the annual rainfall on Titan. Anywhere on Earth with such low rainfall would qualify as a desert.

However, living on a hydrologically active planet should warn us to be wary of thinking only about 'average' conditions. Typically, the desert region around Tucson in the south-western USA receives a mere 30 cm of rain in one year, considerably less than the global average. Even so, the landscape of the area is sculpted by water erosion, partly because the rain falls in sudden, powerful downpours.

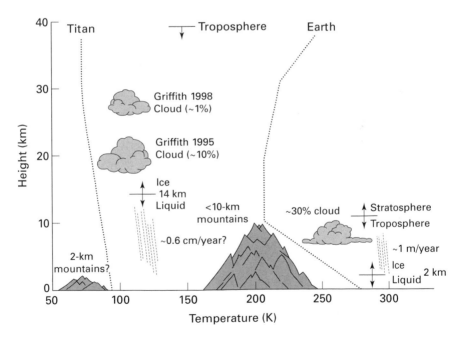

Figure 4.3. A comparison between the tropospheres (the parts of their atmospheres nearest the surface) of Titan and Earth. Solid lines show variation in temperature with height. Mountains, clouds and rainfall are depicted. Adapted from a diagram by Ralph Lorenz published in *Science*, vol. **290**, p. 467 (2000).

109

Titan may be similar. Rain cannot be frequent, otherwise the supersaturation hinted at by *Voyager* could not be sustained. But it may be spectacular. The sheer amount of methane in the dense Titan troposphere implies that a single event could dump a meter of rainfall. While most rivers on Titan may be dry, river valleys may yet be abundant and deep.

RALPH'S LOG. 1993.

I first began thinking about methane rain on Titan in a slightly different context. During the early stages of the Huygens project, the concern arose that droplets of methane might freeze onto the probe and degrade its aerodynamic performance or interfere with its measurements.

In evaluating this icing hazard, one important consideration was how big the drops of methane might be. Tiny fog droplets would just follow the streamlines of the air around the probe and never actually hit it; above a certain size, the drops would have enough inertia to push across the streamlines and hit the probe. If the drops were supercooled, they'd freeze in place.

I dug up papers and books about raindrops and cloud droplets. Because the only example of rainfall is on Earth and is very familiar, most theoretical work on the subject turns out to be 'empirical': in other words, ad hoc equations to describe the shape and descent speed of drops without particular regard to the atmospheric and gravitational environment in which the rain falls, or the properties of the material from which the raindrops are made. All that the meteorologists cared about was something that worked.

There had been a paper a few years earlier that had treated raindrops as big aerosol particles and had worked out how fast they'd fall by balancing their weight against the drag due to the viscosity of the air. This is an accurate method for very small particles, where viscosity is very important. However, I was coming at the problem from a slightly different perspective, that of aeronautical engineering. I was used to thinking about large objects pushing through the air and knew that raindrops were rather too large for this aerosol approach to work. The mass of gas the droplets had to push out of their way was more important than the gas's viscosity.

I constructed some equations to describe the fall rate of spheres in a given atmosphere using the drag equations that included this scale effect. It worked fine for very small particles and for medium-sized raindrops. But it didn't work for the largest raindrops, which were the most interesting. My sphere model predicted that water drops on Earth would fall rather faster than measurements showed that they do.

Something wasn't right and it took quite a few hours in the library to work out what: big raindrops aren't spheres.

In fact, falling raindrops tend to be rather flattened – quite the opposite of the 'teardrop' shape people tend to assume. Because they are falling quickly, we never get a good look at them in daily life. To see them properly you need to look in a vertical wind tunnel, where a drop can be held steady in an updraught of air. Since the air is moving upwards as fast as the drop is falling through it, the drop stays fixed and can be photographed. Bigger drops fall faster, as the model predicts but, as they fall faster, the aerodynamic forces on the drop distort it, making it flatter and flatter. When the width of a drop is about twice its depth, its centre blows through and the drop breaks up. This occurs because the surface tension of the drop can no longer balance against the aerodynamic forces. (Surface tension is the strength of the 'skin' of a fluid – the property that allows insects to walk on water – and is an intrinsic property of the material.)

Now I was getting somewhere. When I multiplied the drag on the drop by a factor related to the ratio of the aerodynamic forces to the surface tension, the model agreed perfectly with the data for terrestrial raindrops and I could have some confidence in predictions for Titan.

Once the model worked, it was easy to switch the density and viscosity of air with those for Titan's atmosphere, Earth's gravity with Titan's, and the density and surface tension of water for those of liquid methane. Out popped the result: the largest drops that could exist on Titan would be 9.5 mm in diameter, compared with about 6 mm for raindrops on Earth. What's more, these drops would fall very slowly, at 1.6 m/s (about the speed a ping-pong ball falls when dropped on Earth) compared with 9 m/s for raindrops on Earth. The physics was sound and – although perhaps unscientific, important nonetheless – the result also 'felt' right.

Apparently it felt right to science fiction author Stephen Baxter as well who wrote in his sci-fi story Titan,

> The biggest drops were blobs of liquid a half-inch across. They came down surrounded by a mist of much smaller drops. The drops fell slowly, perhaps five or six feet a second. It was more like being caught in a snowstorm, with the flakes replaced by these big globules of methane liquid. The drops weren't spheres, they were visibly deformed into flat hockey-puck shapes, flattened out, she supposed, by air resistance . . .

The speed at which the drops fell didn't substantially affect the icing problem, since the number of droplets would be more important than *the size of them. Still, it was a fun little problem to explore. But would these drops ever make it to the ground? The Voyager radio-occultation data showed that the lowest few kilometres of Titan's atmosphere were not methane-saturated, so that raindrops would evaporate as they fell. On Earth this phenomenon is called virga. It isn't particularly common in wet northern Europe but, in the dry south-west of the USA, it is not an unusual sight and, to parched observers on the ground, a frustrating one as grey tendrils of rain reach valiantly for the ground. But Titan's drops fall so slowly that even the biggest would have time to evaporate before they hit the ground.*

Baxter again:

The rain starts by nucleating around particles in the upper atmosphere. That stuff is usually suspended higher up, and won't reach the surface. But it can be transported down by the weight of the rain, down to lower altitudes. When the rain stops, the last drops evaporate, leaving their cores exposed. The rain ghost.

The term 'rain ghost' may well exist but I had certainly never heard of it before I decided to use it in my paper 'The Life, Death and Afterlife of a Raindrop on Titan' (published in Planetary and Space Science in 1993) to refer to the residual ethane drops that might be left behind after a rainstorm on Titan. It was rather amusing to see my semi-frivolous term adopted in science fiction!

Supersaturation

Yond Same Black Cloud, Yond Huge one, looks like a foul bombard that would shed his liquor.

W. Shakespeare – *The Tempest*

One of the robust and convenient assumptions built into models of Titan's atmosphere has been that the amount of methane is limited by saturation. For a given temperature only so much methane can be present, otherwise it would condense. Near the surface, as on Earth, the atmosphere in general is perhaps only 50% saturated, in part because of the presence of ethane. If the mixing ratio at the surface were 50% of saturation (i.e. about 10 mbar of methane, or 6% by volume) one would expect the rest of the atmosphere to be the same 6% because the methane would diffuse through it. However, because the atmosphere gets cooler higher up, there would be some altitude, perhaps 10 km or so, where this proportion of methane would be all that is needed for saturation. At higher altitudes we would expect the

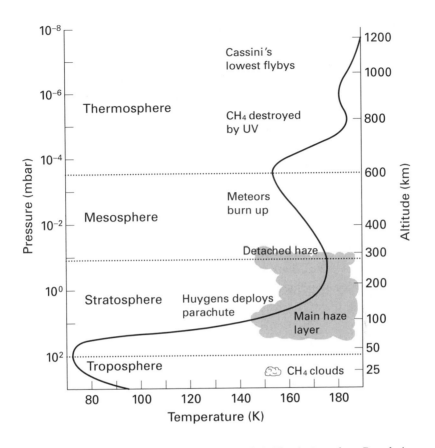

Figure 4.4. The variation of temperature with height in Titan's atmosphere. Boundaries between the major regions of the atmosphere (the troposphere, the stratosphere, the mesophere and the thermosphere) occur at heights of about 40 km, 290 km and 600 km. By contrast, in Earth's much less exteneded atmosphere the equivalent boundaries are at 11 km, 40 km and 85 km above ground. Adapted from a diagram by F. M. Flasar in *Planetary and Space Science*, vol. **46**, p. 1109 (1998).

mixing ratio to follow the amount allowed by saturation up as far as the coldest point. The temperature there is around 70 K and the methane vapour pressure 2 mbar or about 2% of the local pressure. It would be more or less the same from there upwards. The coldest layer acts as a 'cold trap'.

In 1995 the assumption that the amount of methane in Titan's atmosphere was limited by saturation came under attack from Frenchman Régis Courtin and his long-time collaborator Chris McKay. They found that the *Voyager* infrared spectra could be best explained if more methane were in the atmosphere than saturation permitted. Specifically, in the upper troposphere between 10 and 40 km above the surface, it seemed that the methane amount should be about 40% higher than that allowed by saturation.

How could this be? On Earth excess saturation of the atmosphere by water vapour rarely exceeds 1% before clouds form and suck out the excess. Droplets form in milliseconds. When flying on moist days you can see droplets forming an opaque cloud above the wings of aircraft, particularly at takeoff or in tight turns when the high degree of lift requires low pressure (and consequently chilled) air on the upper surface of the wing. But these clouds form because Earth's atmosphere is full of small particles such as smoke and ash from industry and sea-salt aerosols, which act as condensation nuclei on which the droplets can form. It is hard for droplets to form in the absence of such nuclei. Without them, a high level of supersaturation can be sustained. You can demonstrate exactly the same effect with a beer. Open a bottle and a hiss releases the gas above the beer, making the beer itself supersaturated with dissolved carbon dioxide gas. Little trails of bubbles rise in the bottle from small imperfections in the glass which act as nucleation sites. Clink another bottle on top of the first. The shock knocks loose a flurry of new nucleation sites and the beer will froth. On Titan, the nucleation sites are presumably haze particles that slowly settle out of the atmosphere. While the atmosphere is obviously heavily laden with haze, it is opaque not because there are many particles in every cubic metre of air but because the atmosphere is so vertically extended. The actual number of particles falling down is quite small.

In 1997, Bob Samuelson confirmed Courtin's surprising results, using a different analysis of the same data. He also showed that the paradox of supersaturation and rare raindrops made sense. Diffusion would transport methane upwards from the relatively warm saturated ground levels and allow it to accumulate in the colder altitudes where that same mixing ratio would be supersaturated. This upward diffusion would be balanced by methane falling as rain. There was nothing in this model to discriminate between some 'gentle average' Titan where one raindrop fell in one square metre every two weeks and a situation in which violent storms occurred at intervals of months or years.

RALPH'S LOG. 1993–1999.

One reason the idea of rainfall appealed to me is that it offered a means of differentiating the surface. If there were no rain, then the surface should just get covered with a uniform coating of organic muck from the stratosphere and would be all the same colour. Not very appealing. Caitlin Griffith had a similar idea in her 1991 paper and noted that methane rainfall might wash gunk down into low-lying areas.

When I ran my rainfall model, it became apparent to me that this process of highland washing made sense. Because the drops evaporate as they fall, it meant that hills would receive much more rainfall than lowlands. So mountains might have white tops on Titan because the dark organics would be washed away, just as mountains on Earth are white-capped by snow. I even read that mountains on the seabed on Earth also have white tops, as the chalky shells of dead sea-creatures snow onto them but below some depth in the ocean the shells dissolve, so there is a 'snowline' there too. It all made sense – except for a small difficulty, which I ignored, but which gnawed at the back of my mind for several years: much of the organic stuff didn't actually dissolve in liquid methane, so rainfall wouldn't be particularly effective at washing it away.

In the last couple of years, Patrice Coll and others from the LISA laboratory in Créteil (a suburb of Paris) performed some experiments with nitriles, nitrogen-bearing organic molecules that have been detected in Titan's atmosphere. They condense in the lower stratosphere and, importantly, are soluble in hydrocarbons. This helps condensation. If the little nuclei have nitrile coatings, methane should be able to condense onto them quite effectively. Coll's experiments showed that the sort of materials making up Titan's haze dissolve in nitriles. This implies that methane raindrops doped with nitriles should be more effective at washing away organics from elevated terrain. In effect the nitriles are acting as a 'soap', dissolving the otherwise insoluble material, just as real soap dissolves fats that are insoluble in water.

So maybe the rainfall idea has hope yet.

When the HST images came back showing a big bright region, it seemed to cry out 'mountains', although we knew that the edge of the bright region was at more or less the same elevation as the centre of the dark hemisphere of Titan, because these were the two locations at which the *Voyager* radio-occultation measurements were made and they showed the planet's radius to be the same to within half a kilometre.

Athena Coustenis tried a quite ingenious analysis of the near-infrared spectra to see whether the bright region could be explained by altitude alone. Since there is some absorption by methane even at wavelengths where there are methane 'windows', a mountain would seem brighter because it poked up above much of the absorbing gas. Unfortunately (or fortunately, if you're a fan of rainfall!) the mechanism didn't seem to be realistic. A mountain would have to be an implausible 10 km or more tall. This seems unreasonably high for several reasons. First, a mountainous region that high would want to swing around under tidal forces to be on the sub- or anti-Saturn side of Titan, much as the Moon's long axis points towards Earth. Secondly, with Titan's internal heat flow, the ice at 10 km depth would be too soft to support the mountain above it, like an ice cream mountain on a hot plate. Another possibility Coustenis looked into was frost deposits, since the mountaintops above a few kilometres would be colder than the freezing point of methane. This effect allows mountains to be smaller but still bright enough. However, mountains several kilometres high are still needed. More importantly, the antifreeze effect of nitrogen in methane – as we presently understand it – means that in fact the mountaintops would have to be many kilometres higher after all in order to allow frost or snow to form.

Colours of the rainbow

Rainbows are the spectacularly beautiful result of water droplets in the air reflecting sunlight. After light has entered a raindrop, it is reflected from the drop's internal surface before re-emerging. Different wavelengths of light follow slightly different paths as they cross the boundary between air and water and back again. Each colour comes out of the drop at its own angle. Consequently, there is a spread of about two degrees between red and blue light, while overall a primary rainbow forms part of a circle with a radius of about 40°. These numbers are particular to water droplets. Other liquids would not necessarily produce rainbows of the same size. It all depends on the amount by which light veers from one course to another when it crosses in and out of a particular material. This property of a material is summed up in a quantity known as its refractive index. For water its value is around 1.34, varying slightly according to the wavelength of the light.

So would we see rainbows on Titan, as feeble sunlight plays on clouds of methane droplets? If so, what would they be like? First, any rainbow would be larger than on Earth because the refractive index of methane is lower than that of water. For methane at 94 K with a dash of nitrogen added, the refractive index is around 1.27. The rainbow would be about 52° across. On the other hand, higher in the atmosphere there might be ethane-rich droplets. Ethane has a refractive index at 120 K of about 1.31, so the rainbows would look more the size of ours at around 45°. But Titan's rainbows would be much less colourful than we are used to. Since only near-infrared and light from the red end of the visible spectrum make it through the haze, they will be sombre affairs, with no violet or blue or green, and only a hint of yellow. But they will have a special feature of their own – dark stripes corresponding to colours in sunlight absorbed by methane.

Everyone is familiar with rainbows but there are many other less well-known phenomena that produce illuminated displays of one kind and another in Earth's atmosphere. Ice crystals are responsible for many of them. Parhelia, popularly called 'Sun dogs', are one of the more common. Seen as bright patches with some rainbow colours against thin high cloud, they lie 22° or so away from the Sun, one on either side. These patches are the result of light reflected by the 60° facets of flat ice crystals high in the atmosphere. If some of the curious compounds in Titan's atmosphere condense into crystalline shapes, there may be Sun dogs or other optical phenomena. The *Huygens* probe will have to keep its eyes peeled.

Lightning

Rain there may be but is it possible that lightning flashes through Titan's hazy skies? The answer is probably not. Lightning has been seen on Jupiter and crackling radio emissions signalling electrical storms have been detected from Saturn and Uranus. But when *Voyager* 'listened' for radio emissions from lightning on Titan, it detected none. To appreciate why thunder and lightening are unlikely on Titan it's helpful to examine how they happen on Earth.

To get the right conditions for lightning, huge concentrations of positive and negative electrical charge have to be separated out in the clouds. No-one understands exactly how it works but small drops and ice crystals tend to become positively charged while larger drops and hailstones acquire negative charge. Large drops fall more quickly and can overcome the updraught in a storm cloud to sink to the bottom, while small drops get carried to the top of the cloud. The result is a cloud with a positively charged top and a negatively charged bottom. As the cloud builds, so do the charges, until the air can no longer stand the tension. A large spark arcs across the cloud, to another cloud, or to the ground, to equalise the voltages. This spark is the lightning bolt. A current of some tens of thousands of ampères flows for a few milliseconds and heats the air to around 30 000 K. This white-hot thread of air expands rapidly, making the thunder clap. The lightning current creates a strong radio pulse over a wide range of wavelengths. You can sometimes hear this as a crackle on the radio.

Now what about the situation on Titan? There is very little convective heat flow to generate clouds and separate electrical charges. What's more, methane is quite different from water in its electrical properties. Water molecules are polar. That means that the electric charge of each molecule is unbalanced, with a positive end and a negative end. By contrast, methane molecules are not polar. Water is able to sustain higher electrical fields and has stronger surface tension than methane. If a liquid drop is charged more and more, eventually the electrical field will overcome the surface tension and tear the drop apart. Methane drops can hold much less charge than water drops before they shatter. It may be that the polar nature of water molecules is important in generating the charges in the first place too. All in all the prospects for lightning on Titan do not look favourable

RALPH'S LOG. EARLY 1991.

In the project team at ESA, I was the young guy straight out of university among veteran engineers and managers with decades of experience. This meant I got plenty of menial tasks, like checking one document for compliance with another, but it also meant I got some interesting jobs because they didn't fall into anyone else's brief. After all, Huygens was not a satellite, so there were many questions about Titan's environment, none of which were familiar to designers of 'regular' satellites.

One question was that of lightning. What was the risk that a lightning strike could hit the probe? Where contracts are involved, there has to be a specification. And if there is a specification that the probe should continue to operate after a 20-kA lightning strike, there would have to be a test to demonstrate that. So these questions assumed serious importance. I trotted off to the library to learn about lightning. I found out that US missile systems have a specification that requires them to survive nearby but not direct strikes and also that airliners are struck about every 3000 flying hours.

Through my short study, which explored how much energy was available for the convective motions that generate lightning and how much charge an individual drop of methane could hold, it looked as if a lightning strike with a probability of occurring during the probe mission of one in a thousand should have a lower charge associated with it than a typical terrestrial strike. The specification on the probe was relaxed to 4 kA.

It was still a big deal to test the probe's systems for 4 kA. An additional psychological aspect was that many of the ESA Huygens team had worked on a mission called Hipparcos, which mapped star positions from orbit. Although, it performed an excellent scientific mission, it did so only by the skin of its teeth, having been stranded in a low orbit because its main engine had refused to fire. One speculation had been that a lightning strike shortly before launch might have damaged the firing circuit.

The Huygens tests used a huge voltage generator, a bit like the Van de Graaf generator used in classroom physics demonstrations. The team held their breath as strikes were made to different parts of the probe. None of them had ever seen this kind of test – most satellites don't need them. A couple of the strikes caused one of the two telemetry channels to lose a few seconds of data. The most sensitive test, close to where there was a lot of cabling in the probe, caused both channels to lose six seconds of data. Still, that was why Huygens had two links, with the data staggered by six seconds between them, so that the overlap would allow recovery from just such an event. On the other tests the hardy probe survived without skipping a beat and detailed inspection afterwards showed that the probe had not sustained permanent damage. The support electronics used to monitor the probe during the test, however, showed how harsh the tests had been. Even though this gear was twenty feet away and not connected to the probe by any wires, during one test it locked up completely and had to be reset by hand!

The lone voice in support of lightning was ESA's Réjean Grard. He argued that perhaps an ionized layer created by meteors had shielded *Voyager* from the lightning emissions. His instrument on the *Huygens* probe will measure electrical fields and waves in the atmosphere and the electrical properties of the surface. It will be able to detect emissions from lightning, which may account for his enthusiasm. It has been pointed out, however, that the *Voyager* radio occultation experiment would have detected such a layer.

RALPH'S LOG. OCTOBER 2000.

At a Mars Terraforming conference organized at NASA's Ames Research Center by Chris McKay, I met a Mexican physicist whose name I had seen in connection with recent experiments on simulated lightning in Titan's atmosphere. I was quite blunt and asked, 'Why bother?' Rafael Navarro-González explained that the amount of ethene in Titan's atmosphere was rather higher than predicted by the photochemical models and ethene is efficiently produced in lightning discharges (or their weaker cousins 'corona' discharges). While my confidence in photochemical models is far from absolute, this was a fair point. Lightning might be the mechanism for generating excess ethene. Rafael also spoke of an old Russian paper on the synthesis of organic compounds by discharges in hydrogen cyanide. It mentioned that the hydrogen cyanide became electrically charged as it froze. This, or other as yet unknown mechanisms, might produce the charge separation needed for lightning.

Driving winds

The ultimate driving force behind winds in planetary atmospheres is sunshine. On a planet that is not rotating and has a more or less transparent atmosphere, the winds would simply blow from the subsolar point to the opposite side of the globe. In other words, the winds would transport heat from the hottest part of the planet to the coldest part. The flow would look like the field lines of a bar magnet.

Rotation complicates matters. Its first effect is to distribute sunlight all around the planet. Assuming the Sun is overhead at the equator, the greatest amount of solar energy is deposited around the equator and

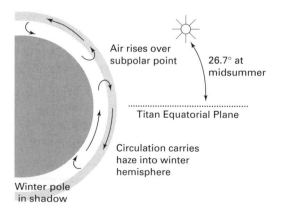

Air rises over subpolar point

26.7° at midsummer

Titan Equatorial Plane

Circulation carries haze into winter hemisphere

Winter pole in shadow

Figure 4.5. A schematic diagram of atmospheric circulation on Titan of the 'Hadley cell' type. Warm air rises over the summer hemisphere blowing haze over into the winter hemisphere.

the least at the poles, with a gradual fall-off in between. Again in this situation, the winds act to remedy the temperature imbalance. Over the hot equatorial regions, the air heats up and rises (warm air being less dense than cold air), then flows towards the polar regions where it descends. This kind of circulation is called a 'Hadley cell' (illustrated in Figure 4.5). On Earth, there are actually several cells between the equator and the poles rather than one huge circulation.

But there's more to it. Our planet rotates as a solid body with a constant angular velocity. As a parcel of air moves away from the equator, its distance from Earth's axis of rotation gets smaller. To conserve angular momentum (the product of mass, radius and angular velocity), the parcel of air moves faster. The effect, called the Coriolis effect, is to make the poleward flow swerve forwards. This gives rise to the trade winds. Similarly, the descending air flowing from high latitudes to lower ones to complete the Hadley cell are diverted backwards. Coriolis effects lead to the belts in Jupiter's atmosphere. Followed over several days or weeks from a satellite, Earth's atmospheric circulation shows up as similarly banded. The faster a planet rotates, and the greater its diameter, the larger these effects are. On a slowly rotating planet, the Coriolis effects are small and the banding is more subdued.

Another important factor determining wind patterns is the altitude at which sunlight is absorbed. If the sunlight is absorbed above most of the atmosphere, it follows that most of the atmosphere will not be subject to vigorous equator-to-pole circulation. Furthermore, an upper atmosphere warmed by solar energy discourages vertical motion. Air near the surface won't rise unless the air above it is cooler.

Opaque atmospheres around slowly rotating bodies seem to lead to zonal winds following lines of latitude. Our sister planet, Venus, is the principal example. By analogy with Venus, Titan's atmospheric circulation was expected to be dominantly zonal, with Hadley cells playing a very small part, a theory supported by observations from the *Voyager* infrared spectrometer. The wind speed should be around 100 m/s at an altitude of 200 km or so. Confirmation came through an amazing fluke of nature.

Thanks to a lucky star

On the 3rd of July 1989, astronomers were standing by for what would be an incredibly fortuitous event. It had been known for several years

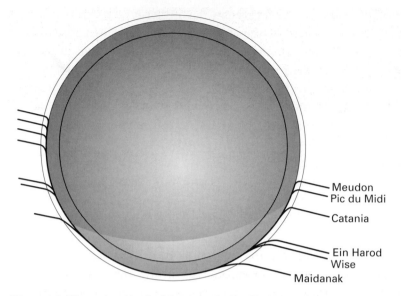

Figure 4.6. The apparent track of the star 28 Sagittarii relative to Titan as seen from various observatories before and after its occultation by Titan on the 3rd of July 1989. The star's image appeared to be deflected around the edge of Titan because of refraction in the atmosphere. The shading on Titan represents the haze thickness. Adapted from a diagram by W. B. Hubbard in *Astronomy and Astrophysics*, vol. **269**, p. 541 (1993).

that a star called 28 Sagittarii would pass behind Saturn and its rings that night. The star was relatively bright at magnitude 5.5 – bright enough to be seen by the naked eye in a truly dark sky. Early in 1989 it had been suggested that Titan might cross in front of the star too. This was an amazing piece of luck. The occultation would provide a unique opportunity to probe Titan's atmosphere from Earth. On average, such an event would be expected only once every few thousand years. Larry H. Wasserman of the Lowell Observatory predicted an occultation visible from Europe. But there were considerable uncertainties inherent in the calculations. No-one could be confident about what would actually be seen from any particular place. Observers were at the ready at many locations along and around the predicted track of Titan's shadow, including amateurs in Britain and a team from Arizona with a mobile telescope set up on an Israeli kibbutz. There was a good chance of at least some of them having clear skies.

Because of Titan's atmosphere, the starlight would not be cut off abruptly. But it wasn't expected to fade out smoothly either. Rather, a series of flashes or 'spikes' were anticipated. Spikes had been seen when stars were occulted by Jupiter in 1971 and by Neptune in 1968.

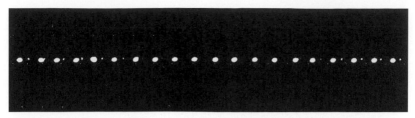

Figure 4.7. A multiple-exposure photographic record of the occultation of the star 28 Sagittarii by Titan on the 3rd of July 1989 made by a British amateur astronomer, Harold Ridley. The 5-s exposures were taken approximately 30 s apart with a 170-mm (6.8-inch) refractor. The large oval dots are overexposed images of Saturn. 28 Sagittarii, to the upper right of Saturn, is absent from the images in the middle of sequence when the occultation by Titan was taking place. Faint images of Titan are visible on the original negative but do not show up on the high contrast print reproduced here.

They result from the way in which starlight is deflected from a straight path when it passes through an atmosphere in which there are gradations of density and temperature. The angle by which the light path changes is determined by the refractive index of the gas. Its refractive index depends on its density, which in turn is affected by altitude and temperature. Anyone who has driven on a sunny summer day has probably seen this effect at work. Mirages – what look like pools of water on the road ahead – are light from the sky forced to follow a curved path rather than a straight line because of the temperature gradient in the air over the hot road. Light from a star behind the edge of a planet can be bent right around the edge by the atmosphere. Predictions of what would be seen based on various models of the conditions throughout Titan's atmosphere could be compared with the observations to find out which fitted best. That was the hope anyway as astronomers waited anxiously just after 10.30 p.m. Universal Time.

For once, astronomers found their luck was in. The prediction of when and where the occultation would be visible was pretty accurate and skies were clear. The event was recorded from as far north as Sweden to the Pic du Midi Observatory in the Pyrenees mountains of southern France. The starlight fell off over about 20 s as expected, complete with 'spikes', but then observers were in for a total surprise. The occultation was expected to last some five minutes. About half way through, there was a bright flash as if the star were shining through Titan! It lasted around 5 s at Paris and a shorter time elsewhere. In fact, what happened was that, as the star got close to being directly behind the centre of Titan, its light could be bent around the entire edge of Titan, as if Titan's atmosphere were a lens focussing the star's light onto Earth.

The shape and timing of the central flash, recorded at several stations, gave a bonus of yet more information about the atmosphere. The number of observing stations able to see it was astonishingly fortuitous. Although occultations by Titan of such a bright star would be expected once every 50 000 years, seeing such a central flash would be expected only once every million years!

The wealth of data was still being analysed more than a decade later. Among the many results gleaned from it have been some things about the nature of Titan's winds. First, the atmosphere is somewhat flattened at the poles. The surface defined by a pressure of 250 mbar (at roughly 250 km altitude) was around 30 km higher over the equator than at high latitudes. This is consistent with the upper atmosphere rotating quickly. Saturn itself is visibly squashed, spun out into a flattened sphere by its rapid rotation. Then the data helped measure not only the zonal winds but also the distribution of haze in the atmos-

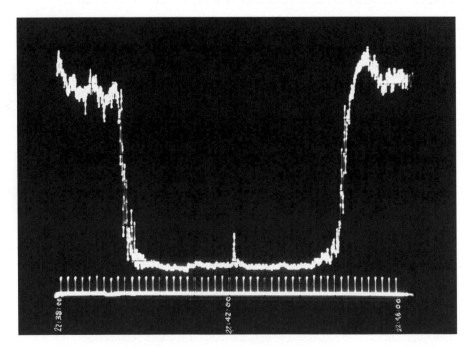

Figure 4.8. A digitised video trace of the brightness of the star 28 Sagittarii during its occultation by Titan on the 3rd of July 1989 made by British amateur observer Terry Platt. It shows a series of 'spikes' at the beginning and end of the occultation as well as a central spike. The spikes are a result of refraction by Titan's atmosphere. They were seen by visual observers as a series of flashes.

phere, which was thicker in the northern hemisphere than in the south, and temperatures at heights between 250 and 500 km – rather higher than the layers probed by the *Voyager* radio occultation experiment.

The stellar occultation gave reassuring confirmation of the theoretical predictions of fast winds at high altitudes. But still there was no information on their direction.

Which way does the wind blow?

In the early 1990s, scientists began to apply general circulation models (GCMs) to Titan's atmosphere. These are giant computer programs, often requiring supercomputers to run them, which divide up a planet's atmosphere into a three-dimensional grid and calculate the pressure changes and flow speeds generated by the input of solar energy. Highly developed models of this kind are used to predict the weather on Earth. By changing some of the parameters, like planetary radius, gravity, pressure and so on, the models for Earth can be adapted to other atmospheres. Although it took considerable fiddling with the models to make them generate a super-rotation of the magnitude indicated by the *Voyager* and stellar occultation data, they consistently produced winds blowing in the prograde (forward) direction. Certainly, the atmosphere is so massive that if it is observed to spin in one direction one year, there is no way it could slow down and then spin up in the other direction in a short time. But should the *Huygens* probe's mission be staked on the 'expectations' of atmospheric scientists and the fiddle-sensitive results of an inscrutable computer program?

In 1995–7, a delicate but powerful technique was applied to resolve the question of the rotation direction. As with so many other questions, it was spectroscopy that gave the definitive answer. The method required observations of very narrow fine lines. There are indeed many fine emission lines in Titan's spectrum. They result from energy transitions in particular molecules and occur at well-determined wavelengths. To pick out a fine emission line means spreading out the light very thinly. The larger the number of wavelength bins the light is divided into, the less light in each bin. The strength of the signal becomes critically low, especially as Titan is faint anyway at the wavelengths where these lines are found.

Ted Kostiuk of NASA's Goddard Space Flight Center used the 10-m

Keck I telescope to measure the Doppler shift of an ethane emission line from Titan. The Doppler shift acts as a celestial speedometer, the change in wavelength from normal being in proportion to velocity. The Keck was able to garner a just-about-adequate signal and its large aperture translated into a small field of view – about the size of Titan itself. Kostiuk and his team pointed the Keck separately to the east and west sides of Titan. They found that the wavelength of the ethane line changed between the two observations just as it should if the atmosphere on one side of Titan were coming towards Earth and moving away on the other. Moreover, the amount of the shift was around that expected for winds of approximately 100 m/s.

As was generally expected, it seems from this observation that the atmosphere rotates in a prograde sense. While no-one was especially surprised, it was something of a relief to both *Huygens* scientists and to theorists who would be hard-pressed to explain an atmosphere rotating backwards. *Huygens*, as designed, can cope with winds blowing in either direction but there will be a better safety margin on the relay link if the antenna can be pointed to where the winds will blow it, a consideration that will be particularly important at the end of the mission.

During the *Huygens* mission, it will be possible to measure the winds by monitoring the Doppler shift of the probe's radio signal as received by the *Cassini* orbiter and subtracting the known Doppler shifts due to the motion of the orbiter and the descent of the probe. The contribution to the Doppler shift that is left will be due to the east–west and north–south motion of the probe. The Doppler measurement will be so sensitive, that even the slow rotation of the probe, whose antennae are displaced slightly from its central axis, may be detectable.

It may turn out that our picture of zonal winds is overly simplistic. Perhaps there are giant weather systems associated with methane condensation that swirl around, as on Earth. While it seems most likely that Titan is simply too small to experience the kind of cyclonic weather systems we have on Earth, the solar system has surprised us before. For example, from the images returned by the *Voyager 2* spacecraft as it followed its grand tour to the outer planets, it began to look as if planetary atmospheres were more bland and placid the farther they were from the Sun. First was turbulent Jupiter, then gently banded Saturn. Uranus seemed duller still, with hardly any features in its atmosphere at all. But then Neptune, with even less sunlight available to drive winds than Uranus receives, shocked everyone with its superfast winds, giant dark spot and white clouds. The only way to find out is to look. As well as *Huygens*' making direct measurements, the *Cassini* orbiter will be able to track cloud motions using filters that probe down deep into the atmosphere.

The gustiness of the atmosphere was another important question for the designers of the *Huygens* probe. If winds change too quickly, the probe might swing under its parachute, mispointing its antenna so that the radio link would be intermittently broken. The best data on this problem comes from the scintillations in the radio and stellar occultations . Certainly the atmosphere should be less blustery than on Earth but beyond that we simply don't have enough information to make a robust estimate.

Winds are in general most gentle near a planet's surface, where drag saps their energy. This, after all, is why windmills are put on towers. Understanding Titan's near-surface winds will be important for future exploration, which might use balloons or helicopters. Winds are also crucial for several processes that shape the landscape, such as the formation of sand-dunes. Waves on lakes and seas are whipped up by the winds too. Would these be wild and violent or is Titan's lower atmosphere like the stagnant depths of a muddy pool? Observations of the upper atmosphere's prograde super-rotation do not help us here. In terms of direction, the eastward and westward winds near the surface must approximately average out, otherwise there would be a net force on Titan as a whole acting to make it spin slower or faster. We are left for the time being to puzzle over models and analogies.

Climate

As inhabitants of planet Earth, we are familiar with the contrast between temperatures at low and high latitudes. The situation is similar on Mars and Titan. The axes of all three of these bodies are tilted by a modest angle in the 23–27° range. On these worlds, high latitude regions get roughly half the amount of sunlight received by equatorial regions. The daily averages vary over the course of a year, by a small amount at the equator and more strongly near the poles, which experience the extremes of a long polar night of winter and the midnight Sun in summer. However, averaged over a whole orbit around the Sun, the rule of thumb of one half holds fairly true.

On an airless, ocean-free world, this uneven sunlight would make a very considerable difference between high-latitude and low-latitude temperatures. We can make a rough estimate of what that difference would be. In equilibrium, energy lost equals energy input. Since outgoing energy flux is proportional to the fourth power of temperature (measured on the absolute or Kelvin scale), the half-lit high latitudes should have a mean temperature lower than that at the equator by a factor equal to the fourth root of a half, which is 84%. Taking the equatorial temperature on Earth as 300 K, or 27 °C, that would put the average high latitude temperatures, covering everything polewards of 30° (roughly north of the Mexican border, all of Europe, Australia and South Africa too) at a chilly 252 K or −21 °C. Very high latitudes would have temperatures even lower than this!

Happily we do not live on such a world. The motions of the air and the oceans act to even out the unfair distribution of sunlight and lower the temperature difference between low and high latitudes. Across the 45° lines of latitude on Earth, around 4 billion megawatts of heat, equivalent to the output of a million large power stations, is transported in warm air, water vapour and warm ocean water. About one third of the total heat transport is due to the oceans.

Understanding the nuances of Earth's climate is mired in detail and something that will take decades yet to understand adequately but, during the 1970s, climatologists were still addressing big questions, specifically the possibility that we could be on the threshold of a new ice age, with models that could be handled with pencil and paper. These models divided Earth up into latitude bands and balanced the flow of energy in and out. Since Earth isn't growing appreciably hotter (forget the greenhouse effect for a moment!) the sunlight absorbed by each latitude band must equal the heat radiated back into space plus the heat transported to other latitude bands. This latitudinal heat transport would be represented in the models by some parameter (usually called 'D' because it is analogous to a diffusion coefficient in other physical problems) multiplied by the difference in temperature between the bands. The greater the temperature difference between neighbouring bands, the more heat flows in the polewards direction. For Earth, modellers just picked a number for D that works to produce the latitudinal temperatures we see today: about 300 K (27 °C) for low latitudes, falling smoothly through 285 K (15 °C) at 45°, to 250 K (−23 °C) or less at the poles.

If we take the value of D determined for Earth and apply it to Titan (changing other aspects like distance from the Sun, tilt of axis and the greenhouse effect), we would expect Titan's high- and low-latitude regions to have the same temperature to within a few hundredths of a degree. However, *Voyager* infrared measurements that should tell us the temperature of the air near the surface say that the temperature actually falls off by around 4 degrees at high latitudes. What is going on? Using the same value for D should, if anything, over-estimate the temperature contrast, since Titan is smaller and has a thicker atmosphere than Earth.

One possibility is that the temperature measurements are wrong. Perhaps the data is reflecting temperatures higher in the atmosphere, which are known to be lower at high latitudes from measurements at other wavelengths. The other possibility is that the meteorology isn't well understood. Something is happening on Titan to make heat transport less efficient than it is on Earth. Perhaps the zonal winds disrupt the Hadley circulation somehow. As suggested by David Stevenson, a New-Zealand-born professor at Caltech, condensation and evaporation of liquid methane and nitrogen near Titan's poles may fix the temperatures there. The same kind of thing happens on Mars, where temperatures never fall below about 150 K, even during the long polar night, because carbon dioxide frost forms. As it condenses, it gives up heat, which balances the loss of heat to space.

Whatever is happening, *Cassini* should provide the data to work it out, using its radar as a radio thermometer, as well as better infrared observations and radio occultations at different latitudes.

RALPH'S LOG. DECEMBER 1999.

The fall conference of the American Geophysical Union is a large, annual meeting held in San Francisco a few weeks before Christmas. In 1999 I walked around the posters and in and out of talks in a bit of a daze. We were supposed to have been presenting exciting new results from the Mars Polar Lander and the Deep Space 2 penetrators, two tiny (2-kg) probes that would bury themselves half a metre down in the Martian soil, near the edge of the southern polar cap. The previous week, instead of landings followed by a triumphant barrage of data and pictures, there was nothing. All three spacecraft were lost without trace. It was a huge blow, although more for others than for me, since over half of my time was spent working on Cassini anyway. Still, I had been expecting to be busy with new Mars data for a few months, and the conference gave me a chance to look for other things to do.

I came across a rather obscure Japanese poster describing a model of the circulation in Earth's oceans, which used a principle suggested by an Australian, Garth Paltridge, back in 1975. I had come across this before but thought it had been forgotten, either ignored as an irrelevant coincidence or discredited altogether. The principle held that the heat flows across Earth's surface happen to be just at the value that maximises the production of entropy, or to a close approximation, that produces the maximum amount of mechanical energy. If the heat transport were too small, there would not be enough heat to drive the atmospheric motions; if too much heat were transported, the temperature difference between low and high latitudes would be small, and the conversion of all that heat into motion would be inefficient. (Although a bath of lukewarm water holds more heat than a bucket of hot water, the latter is more useful.)

This was a very interesting idea. It gave the value of D that agrees with Earth's climate, without assuming anything about Earth's oceans or clouds or any of the other empirical fudges applied to general circulation models. Maybe it would be general enough to apply to Titan.

I played with a little model of Titan's temperatures. I tried using the large values of D that you would expect from Titan's small size and dense atmosphere but the model just wouldn't settle down. Temperatures bounced all over the place. It worked okay with very small values of D but that couldn't be right.

Then I noticed that a small value of D – a hundred times smaller than for Earth – also gave the temperature contrast that had been observed in the Voyager data. I decided to look at the entropy production and, sure enough, it had a maximum value for that very same D. It struck me immediately that this could have profound significance. By studying Titan, it was obvious that Earth's agreement with a maximum entropy production state was unlikely to be a coincidence. This principle would be enormously useful in predicting the climates of Earth and Mars earlier in the solar system's history and those of planets around other stars.

Working with Jonathan Lunine, Chris McKay and Ph.D. student Paul Withers, I developed the idea a little further. Many colleagues – especially those interested in extrasolar planets – greeted the idea enthusiastically and our paper was accepted for publication in Nature, one of the most prestigious journals. Others thought it too good to be true, particularly traditional meteorologists. One scientist invited to comment on our paper lobbied successfully to have it rejected – almost unheard of when a paper has been formally accepted. To arouse such passions in the ranks of scientists suggested we were on to something important . . .

Tedious seasons

Earth's equator is tilted to its orbit by 23.5°. Were this angle (called the 'obliquity') zero, or very small, there would be no seasons. As it is, temperate and high latitudes experience very considerable seasonal swings. Mars, with an obliquity marginally larger than Earth's at 25°, is similarly subject to seasons. In Mars's case, though, there is an extra factor bringing its influence to bear on seasonal fluctuations.

Mars's orbit is significantly elliptical. This means that its distance from the Sun swings between 207 and 249 million km over the course of a martian year. Its speed in orbit varies too, being greatest when Mars is nearest the Sun. Mars' changing distance from the Sun has a profound effect, causing the illumination to vary by 40%. The sunlight is strongest during the northern summer, but, because Mars is speeding around its orbit more quickly at this time, the season is short. Although the total amount of solar energy received by northern and southern hemispheres is the same on average, northern summers are short and intense while southern summers are cooler but longer. This difference may be responsible for the diffent size of Mars' two polar caps. In contrast, Earth's orbit is only very slightly elliptical and our change in distance from the Sun (146 to 150 million km) has a barely noticeable effect. When we are closest to the Sun in January, sunlight is only 2% or so more intense than average.

A planet's obliquity and the eccentricity of its orbit are not absolutely fixed for all time. Slight gravitational perturbations from the other planets cause both quantities to vary. This affects the level of sunlight at different latitudes over the year, which in turn determines temperatures and snowfall and ultimately the growth and recession of ice sheets and glaciers. Such astronomical changes are largely responsible for the ice ages on Earth, as first suggested by Scottish geologist Sir James Croll and later elucidated by Serbian mathematician Milutin Milankovitch.

So what of Titan's seasons? Titan's equator lies very close to its orbital plane, which is a mere 0.3° from Saturn's ring plane. The ring plane is inclined at 26° to Saturn's orbit around the Sun. In effect then, Titan's 'obliquity' for determining seasons is 26°. Perhaps because the saturnian year is quite long (29 Earth years), the moderate eccentricity of Saturn's orbit is often forgotten. It is sufficient to change the strength of sunlight by some 20%, perhaps enough to make Titan's seasons asymmetric like those of Mars. On these grounds, then,

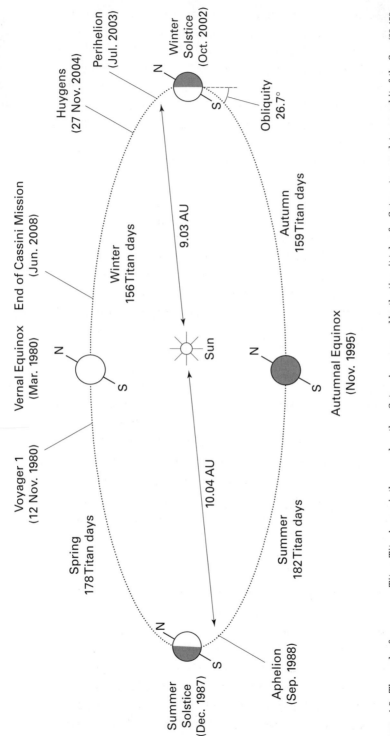

Figure 4.9. The cycle of seasons on Titan. Titan's year is the same length as Saturn's – governed by the time it takes for Saturn to complete one orbit of the Sun (29.458 Earth years, or 10 759.5 Earth days). For clarity, Saturn has been omitted from this schematic diagram. Seasons indicated refer to Titan's northern hemisphere. Opposite seasons apply in the southern hemisphere.

Titan experiences noticeable seasonal effects because its equator is inclined to Saturn's orbit by 26.7°, a little greater than Earth's equivalent tilt, which is 23.5°. Titan's seasons are modified to some extent by the fact that Saturn is 1 AU (150 million km) farther from the Sun in northern summer than it is in southern summer. Adapted from a diagram by Tetsuya Tokano in *Planetary and Space Science*, vol. **47**, p. 493 (1999).

Titan's seasons may be fractionally more exaggerated than Earth's, notwithstanding Christiaan Huygens' expectation that, because the seasons on Saturn are long, they would be tedious!

As this book was being completed in 2001, Titan was nearing southern midsummer and the solstice in 2002. We can think of the years 1999 and 2000 as 'November'. Vernal equinox ('March', the start of northern spring), was back in 1980, when *Voyager 1* flew by, and northern summer occurred in 1997. In terms of Titan's year, the *Cassini* mission will run through 'February' and 'March'.

A hazy understanding

The haze pervading Titan's atmosphere is one of its most exotic and important features. Because the atmosphere is so dense and extends so far upwards into space, it takes haze particles a long time to fall down from where they form, hundreds of kilometres above the ground. As a result, the atmosphere contains a very large amount of haze.

We know, roughly, what the haze is made of. With some effort it can be reproduced by laboratory experiments. Several people have tried. They all found that their haze was a huge jumble of many different chemical compounds but, when they are all added together, the basic ingredients are the three elements carbon, nitrogen and hydrogen, roughly in the proportions of 10:10:1 by numbers of atoms. The exact proportions depend on the details of how the experiment was carried out, such as the ratio of nitrogen to methane in the test chamber, the temperature, whether ultraviolet light, electrical discharge or laser-driven explosions are used to break down the molecules and so on. Sagan and Khare's experiments at Cornell yielded 8:13:4; Patrice Coll in Paris got 11:11:1, while Chris McKay found 11:11:2.

One of the first puzzles and, in a sense one that endures, was the question of the size of the haze particles. This is actually a much harder question than it sounds, since it depends on when and where you are talking about. And, as we shall see, even defining what you mean by size isn't a straightforward matter.

The simplest concept of the 'life cycle' of the haze has organic molecules forming high in the stratosphere where sunlight breaks up the methane. The organics build up to the point where they condense into small droplets or crystals of some size – let's say 0.05 microns. These particles fall down under the action of Titan's gravity but descend slower and slower the deeper they get into the dense atmosphere. It follows that the amount of haze in each cubic metre of air increases at lower altitudes: like traffic on a highway, the slower it moves, the denser it's packed (although the cause and effect are different here, of course).

Unfortunately that model doesn't work because the haze would get impenetrably thick. In that scenario, no sunlight would reach Titan's surface – which clearly it does in real life. In a slightly more sophisticated model, the tiny haze particles begin to stick to each other as they become more densely packed and begin to collide more often. All the haze particles are assumed to be spherical but at deeper levels in the atmosphere the particles are of a larger size. Larger particles fall more quickly than small ones. In reality, there will be a distribution of particle sizes at each level in the atmosphere but some models (including McKay's) assume that there is only one size of particle at each level, determined by the size and number of particles drizzling down from the level above. How easily they collide depends on factors like how much electrical charge they hold; heavily charged particles tend to repel each other and have difficulty sticking together.

Particles of different sizes scatter light in different ways and the different number densities and sizes of particles in these models, assumed to be made of something like tholins, did quite a good job of reproducing Titan's spectrum and the way the brightness of Titan's disk fades all around its edge. However, there were some nagging problems.

First, even with particles coalescing on a reasonable scale, the models still predicted more haze than there is. One likely explanation is that some haze particles are 'washed out' of the atmosphere. The way it works is this. As we move deeper into Titan's stratosphere from above, the temperature drops rapidly. Two hundred kilometres up it's around 180 K falling to 70 K at the tropopause, at a height of 40 km. Various compounds produced at high altitudes will reach saturation point as they wend their way downwards through colder and colder air. The compounds will condense onto haze particles, making them larger and heavier, so they are washed down quickly towards the surface. The models typically accounted for this process by expediently declaring there to be no haze particles below some notional altitude.

Secondly, there were data from the *Pioneer 11* spacecraft (before the *Voyagers*, remember). It measured the polarisation of the light reflected by Titan at a range of angles. Theory tells us that polarisation is quite sensitive to particle size. Considering *Pioneer*'s findings alone, Tomasko and Smith drew the unavoidable but rather puzzling conclusion that large particles must be present above small particles in Titan's atmosphere. And that didn't make much physical sense.

In the late 1980s, Peter Smith and Bob West at JPL realised that the problem with the polarisation data might lie with the assumption that the haze particles are spherical. In physics, the mathematics to handle phenomena involving spheres is usually much easier than dealing with objects of any other shapes. 'Assume a spherical cow' runs the joke, mocking a physicist's typical starting point for a complicated problem. It is true that liquid drops tend to assume a near-spherical shape in many situations because of surface tension but snowflakes have a flat crystalline form due to the peculiar properties of water. In reality, lumps of stuff – soil, or smoke, or instant coffee granules – generally have irregular shapes. Although irregular, these shapes have an intrinsic order. The lumps tend to be agglomerations of smaller lumps, which in turn are often assemblages of smaller lumps yet. These shapes, self-similar on different scales, are termed 'fractals'. Computer simulations of clumping particles often produce branching, spidery shapes of this kind, rather than a tight cluster resembling a large drop.

French researchers Michel Cabane and Eric Chassefière, and later Pascal Rannou as well, found that, at the level where haze forms, tiny spherical particles all of one size (termed 'monomer' particles) would collide to form small spheroidal aggregates about one twentieth of a micron across. By the time they had fallen to 200 km around a month later, the particles were typically 10 or 20 monomers across and with an irregular fractal shape. They would then take around 30 years to fall to the lower stratosphere, where condensation on them(probably of HCN) would make them grow heavy and fall rapidly out of the atmosphere. These fractal aggregates sounded physically reasonable and appeared to match the polarisation data better too, since fractal particles act in some ways like small and large spheres at the same time. A large part of Mark Lemmon's Ph.D. dissertation was devoted to this topic. Some years later, Rannou – one of a long succession of postdocs to enjoy a fruitful collaboration with Chris McKay – included the peculiar optics of fractal particles into McKay's widely used computer model.

The crazy hazy days of summer

About the only striking feature of the *Voyager* images of Titan, apart from their basic sameness, was the brightness difference between the northern and southern hemispheres. It seemed to be a seasonal effect. How could the haze vary with season? An obvious idea was that more haze is produced in the hemisphere where it's summer because there is more ultraviolet light from the Sun. However, it takes hundreds of years for the compounds created by these chemical reactions to form haze dense enough to be seen. The time lag would average out any seasonal fluctuation in haze production.

Perhaps the haze particles were for some reason larger in one hemisphere than the other? *Voyager* images ruled that out. They

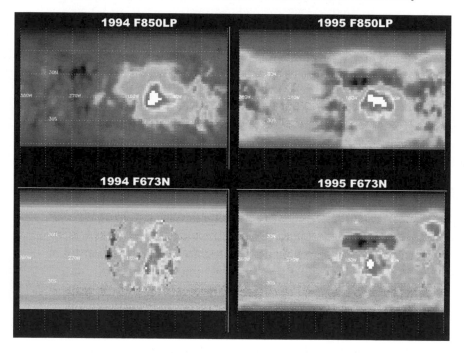

Figure 4.10. A comparison of maps of Titan made from HST images taken in 1994 and 1995 by a group including Mark Lemmon, Ralph Lorenz, Peter Smith and John Caldwell. Images are through two different filters – the F850LP centred around 940 nm (see Fig. 2.8) and the F673N at 673 nm. The large bright feature is apparent in the data for both years. In 1994, the F673N data was only taken around the bright region. The F850LP images primarily sense surface features, while the F673N images barely detect the surface and are quite sensitive to clouds. A bright spot can be seen near the top right of the 1995 maps but it is absent from the 1994 data. Its remarkable brightness near the edge of each image, especially at 673 nm, suggests that this transient feature was a cloud. Image by Ralph Lorenz. (In colour as Plate 8.)

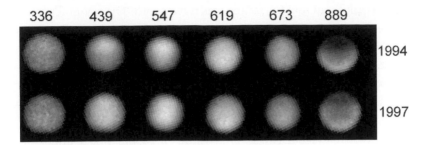

Figure 4.11. Hubble Space Telescope images of Titan taken through a variety of filters in 1994 and 1997. North is at the top and the images have been adjusted so all have the same maximum brightness. The numbers along the top indicate in nanometres the wavelength of light that passes through each filter. They range from the ultraviolet (336 nm) to the infrared (889 nm). Neither of the extremes would be visible to the human eye. The images at 336 nm are virtually featureless. In those at 889 nm, the limb is brighter than the rest of the disk, while at the wavelengths between the two extremes, the disk is darker around the limb than at the centre. The asymmetry between north and south has reversed compared with *Voyager*'s images though it is not as strong in 1997 as in 1994. Image: Ralph Lorenz, published in *Icarus*, vol. **142**, p. 391 (1999).

showed the north–south asymmetry to be about the same whatever the angle of view, which would not happen if it were due to different sized particles.

Progress came with the Hubble Space Telescope even though its resolution was several hundred times poorer than the best *Voyager* images. Hubble's advantage was better signal-to-noise and its data spanned a wider wavelength range than *Voyager*'s. The first HST images of Titan were taken in 1990 by Caldwell and they showed that Titan's north–south asymmetry had reversed. The south was now darker at blue wavelengths, the opposite of what *Voyager* had observed only 8 years before. Caldwell's one image in the near-infrared, in the 889-nm methane band, had the contrast between hemispheres in the opposite sense. It seemed that more haze was present at high altitudes in the winter hemisphere. This extra haze brightened the disk at long wavelengths but darkened it at short wavelengths. The next images came in 1992. The signal-to-noise was better since HST could track Titan and take longer exposures but it was still hindered by faulty optics. The data sets secured in 1994, 1995 and 1997 were better yet, taken with the new sharper camera.

A set of images, taken with the same camera and the same filters over several years is a valuable resource. It was possible to track the change in north–south dissimilarity from one year to the next and it became apparent that the switch was not smooth. The asymmetry flipped quite dramatically from 1995 to 1997 as summer moved into the southern hemisphere. Furthermore, the change was not the same at all wavelengths. The most striking changes over the years were at 889 nm, while Titan stayed fairly constant at red wavelengths with minimal difference between hemispheres. Changes seemed to happen first at the violet and red/infrared ends of the spectrum but later at blue wavelengths in between. This would take some time to understand, since we didn't even know for sure the cause of the asymmetry in the first place, let alone how it changed.

Just increasing the amount of haze produced in one hemisphere didn't work. This made the hemisphere in question too dark in the blue and too bright in the methane band; in other words, there was too much haze at high altitudes. Fiddling with the haze lower down seemed to work better. The quick and easy way to do this was to change the 'rainout' altitude below which there was deemed to be no haze.

Close inspection of the *Voyager* images showed a detached haze layer – thin, but definitely there – separated by a clear gap of 50 km or more from the main haze layer. At the time of the *Voyager* encounter, the detached haze was more prominent in the southern hemisphere and mid-northern latitudes.

Erich Karkoschka, a German-born former student of Tomasko in Arizona, is a meticulous observer of the bodies of the outer solar system. In 1995, they conducted observations of Saturn during the ring-plane crossings, when the rings were edge-on to Earth and all but invisible. To supplement spectroscopic observations of Saturn they

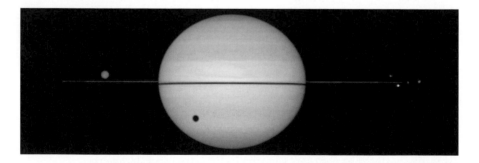

Figure 4.12. A Hubble Space Telescope image of Saturn taken on the 6th of August 1995, when the rings were practically edge-on as seen from Earth. The black line bisecting Saturn is the shadow of the rings. The prominent dark circle on Saturn is Titan's shadow. The four moons clustered around the ring plane on the right are, from left to right: Mimas, Tethys, Janus and Enceladus. Credit: Erich Karkoschka and NASA. (In colour as Plate 9.)

obtained images with the Hubble Space Telescope. Since the ring plane was edge-on to Earth, so was Titan's orbital plane and Titan would transit across Saturn's disk. Cleverly, the HST images were obtained not only with Titan in the field of view but with Titan's shadow cast onto Saturn (see Figure 4.12 and Plate 9). Drawings of this phenomenon have existed for over a century but no useful photographs had been taken during previous ring-plane crossings, which occur only every 15 years.

Titan's shadow, like Titan's disk, was of a different size at different wavelengths: smaller in red light than in blue. Not only that, the shadow was not quite round, almost as if it had a notch bitten out of it. Karkoschka and Lorenz analysed the images and found that they could make sense of this picture if the shape of the shadow were due to the detached haze and it didn't go all the way around Titan. The violet images didn't quite fit but that could well have been because Karkoschka and Lorenz made the simplifying assumption that the particles were spherical, which they knew in reality they weren't. The shadow measurements showed that Titan's limb dipped by some 50 km southward of about $45°$ S. In other words, at 300 km altitude or so, roughly the altitude of the detached haze seen by *Voyager*, the haze was less opaque at high southern latitudes than elsewhere – the opposite of what *Voyager* had seen. These regions were just coming out of winter night. Perhaps some condensing photochemical gases were raining out the detached haze.

Then there were several other bits of evidence to be considered. The stellar occultation data from 1989 showed a slight difference between the altitude of the haze in the north and south. And Pascal Rannou pointed out something that has not received much attention: the dark polar hood in the *Voyager* images seemed to stand up a little higher than the rest of the haze. Not only that, it seemed to connect with the detached haze.

That was certainly interesting. It seemed to fit with an emerging picture of meridional winds blowing haze from one hemisphere to the other. But what did it mean? Was the haze being blown along in the detached layer, to converge and descend over the pole, or was it the other way round? Or is it all purely an effect of temperature? Whatever lies behind the north–south asymmetry is also causing it to change. Although condensation may play a role by making particles grow and sink faster, most researchers now favour the idea of haze being transported by winds from north to south and vice versa.

The first person to explore this quantitatively was Ph.D. student Bill Hutzell, working with Chris McKay. His model didn't quite agree with the observations but it showed the way. To really understand things quantitatively was going to take a brute-force approach. The system is hopelessly tangled: the haze is controlled in part by temperature, which affects the condensation and hence the rainout altitude. The temperature also drives the winds, which move the haze around and control its fall rate. This is already complicated but it gets worse because the haze controls the temperature.

Tetsuya Tokano, a Japanese student working in Cologne, adopted a sophisticated approach. He undertook the monster task of linking a full-blown general circulation model with Chris McKay's haze model. After a couple of years of effort, he got it all working. Although his model atmosphere did not super-rotate as quickly as the observations indicated the real thing does (this is much easier to say than to do), it did capture the essence of the haze cycle. He predicted that the haze would quickly become more opaque to red light as spring turned into summer but would then stay more or less constant. This was good news. It fitted with the HST observations and, as a totally independent piece of information, substantiated the overall picture.

In some respects, the asymmetric haze cycle, with the haze rapidly fleeing to the winter hemisphere like a flock of sun-shy migrant birds, will mean that Titan's haze will be fairly quiescent during the *Cassini* mission: the most obvious seasonal changes high in the atmosphere have already happened in the last few years since the Sun crossed the equator southwards in 1995. Changes in the haze will still continue but, more subtly, at deeper levels in the atmosphere. Not everyone yet accepts this picture as the correct one but, as always in science, time will tell. If it is right, *Cassini* will see rather easier into the southern hemisphere: not only will the Sun be higher over the south but the haze will be thinner.

Light's ups and downs

Wes Lockwood at Lowell Observatory began making systematic measurements of the brightness of several planets in the early 1970s. His simple instrument, a photometer, measured the total amount of light falling on it – like a lightmeter used to set the exposure on a camera. By placing a filter in front of the photometer, the intensity over a small wavelength range can be measured. Lockwood routinely made this measurement for an assortment of objects, looking for variations.

He noted that Neptune seemed to vary in brightness and further-more that a plot of the variation looked suspiciously like a mirror image of a plot of the sunspot number. This suggested that Neptune's atmosphere was influenced by the Sun, probably through its ultravio-let light output, which is known to correlate with sunspot number. Lockwood looked for a similar relationship for Titan.

Titan did indeed vary with time, at both the blue and yellow wave-lengths that Lockwood observed. Blue light varied by about 7% and yellow by 5%, both over periods of about 14.5 years, or half a Titan year. It was only to be expected given the difference in brightness of the two hemispheres, which tilt towards Earth alternately. But it turned out that this seasonal change in presentation could account for only about half of the observed variation in total reflectivity. In the mid-1980s, Lockwood and others found that the 1970–1986 variation in reflectivity could be explained by a combination of the seasonal haze variation and an effect correlated with the 11-year solar cycle. However, by the mid-1990s, the observed variation had fallen out of step with the solar cycle. The agreement over the 1970–1986 period seems to have been a coincidence. With luck a few more years of mod-elling and HST data should work out what is going on.

Lockwood's long and patient record of Titan's brightness has been of great value combined with the better quality, but shorter, record of HST observations. By combining Lockwood's brightness data with measurements of the north–south asymmetry from *Voyager* and HST, we are slowly getting to grips with the changes on Titan. Lockwood's data clearly show, for example, that Titan's yellow brightness changes

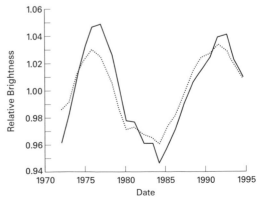

Figure 4.13. Long-term variations in the brightness of Titan as mea-sured by Wes Lockwood of the Lowell Observatory. There is a cycle with a period of about half a Titan year. The cycle for yellow light (dotted line) appears to be ahead of the cycle as measured in blue light (solid line) by about one Earth year.

about half a year before the blue equivalent. Even in these days of space telescopes, the lone, patient astronomer on the mountain with modest, but cared-for, equipment still has a role to play.

Surprise, surprise!

Titan's atmosphere superficially resembles Earth's in many respects and similar kinds of processes occur within it but we have seen that it is at the same time bizarrely different in numerous ways. Titan's distinctive brand of meteorology certainly offers us a new perspective on the perennial topic of the weather. Above all, Titan never ceases to surprise us. And we are constantly finding we have to grapple with the unexpected: clouds after we didn't think there were any, supersaturation after we thought it was impossible, maybe lightning after that too seemed improbable. What will this baffling, magical world conjure up next to challenge our preconceptions?

Titan's landscape

Throughout history, astronomers and visionaries have guessed at the landscapes of other worlds, guided by what little information there is at the time and their imaginations. Venus, shrouded from direct view like Titan, is a parallel for the kind of surprises we can expect when Titan's surface is ultimately revealed.

One early paradigm for Venus was an ocean-covered steam-bath. Another scenario had oily tar everywhere. Eventually, ideas settled on a hot desert. None of the ideas and artists impressions particularly resembled the venusian landscape later revealed by the Russian landers and NASA's *Magellan* radar mapper. Exotic landforms were discovered, unlike anything seen elsewhere in the solar system, such as gridiron terrain and the volcanic blisters called coronae. Familiar structures were seen, including impact craters, volcanoes and wind-blown dust, but Venus put its own twist on these. Many impact craters were surround by broad, flow-like blankets of ejecta and some volcanoes were near-perfect domes, looking for all the world like drops of pancake batter flattened by their own weight. Few dunes were seen, perhaps because there simply isn't much sand on Venus, but wind-blown impact-generated material seems to be responsible for the dark parabolic streaks around recent craters above a certain size.

The development of planetary science has been somewhat like that of zoology. At first we can only catalogue and classify what we see but ultimately patterns and relationships emerge. Planetary surfaces are shaped by physical processes that depend in more-or-less well-understood ways on basic environmental factors, such as distance from the Sun. But the intricate vagaries of the 'real world' interfere with the orderly behaviour that might be expected on the basis of simplified theories. With only nine planets and a few dozen satellites in the zoo to compare and contrast, patterns are obscured by the role chance plays in the lives of planets. Quirks of fate were responsible for the large impacts that gave rise to such curiosities as Earth's giant Moon and the incongruous tilt of Uranus. Nonetheless, over the past 30 years we have had some success in understanding the processes that shape worlds.

It may be a futile exercise to attempt to predict the nature of Titan's surface. Our guesses, however well-guided by sound physics and close looks at the surfaces of other worlds, are certain to be incomplete. There's a good chance that many of them will be utterly wrong because we do not appreciate the key factors. There will be formations and processes we haven't thought of yet. However, speculation about Titan's surface is not wholly without purpose.

First, it gives us an opportunity to test our understanding. It is all very well to explain phenomena and features after *Cassini* or some subsequent mission has revealed them. To predict features *before* they are seen is a rather better test of theory. In this connection, it is worth noting that it is easy, and tempting, to say qualitatively that something is possible – geysers, say. Such predictions can lurk in an obscure paper for decades without being disproved. Rather more solid predictions are null ones – there will be absolutely no glaciers on Titan, for example. The best are quantitative predictions but they are also the most difficult to make, generally relying on calculations involving physical quantities, which in practice may not be adequately known.

Having made some faintly convincing excuses for why what we say might be proved wrong, let's now try and summarise the best guesses about Titan's landscape. The first couple of points are general. One assumption often voiced is that the presence of an atmosphere automatically means there will have been erosion. This is natural enough if a simple comparison is drawn between Earth and the Moon. The Moon has no atmosphere and preserves a virtually pristine record of ancient meteorite bombardment. On Earth, which has an atmosphere, the record of bombardment has largely been obliterated by erosion and resurfacing. A variety of processes wear away rocks on Earth relatively rapidly. This 'weathering' was first quantified by studies of the fading inscriptions on tombstones in eighteenth century Scotland. Rocks are ground down by wind-blown dust, dissolved by acid rain, overgrown and destroyed by plants, broken up by the effects of alternate freezing and thawing, and so on. However, few of these erosive processes are accelerated by higher atmospheric pressure and several are slowed by thicker atmospheres.

Consider erosion by wind. On the one hand, a thicker atmosphere makes it easier for wind to pick up sand and dust. On the other, thicker air acts as a cushion between a blowing grain and the rock it hits. Experiments in wind tunnels at different pressures have shown that the potential erosive power of wind-borne particles decreases as the air pressure is raised. There is also an indirect effect. A thicker atmosphere makes for slower winds. The subtle reason is that winds are a response to the uneven distribution of the Sun's heat. A thick atmosphere can transport the amount of heat needed to even things out by blowing at a lower speed than a thin one would have to. The smaller temperature range experienced by a surface under a thick atmosphere will also moderate the ability of the daily cycle of heating and cooling to break up rocks, a process that is amplified if water in cracks and pores alternately freezes and thaws. The endurance of materials subject to stresses is dramatically reduced when greater deformations are forced on them. You can only flex a paperclip a few times by a centimetre or two but, if you bend it by only a couple of millimetres, you can do so thousands of times. So the modest temperature changes on Titan are unlikely to be destructive. The freeze–thaw cycle of water on Earth is destructive because, when water freezes, it expands, as you can readily demonstrate in the kitchen. Most other materials, like methane on Titan, contract when they freeze, so even if the temperature cycles were deep enough to cause freezing (unlikely) the process wouldn't break up bedrock. All things considered, erosion seems likely to be weak on Titan.

Bull's eyes and horseshoes

Cratering occurs everywhere in the solar system. Those planetary surfaces that have not been altered by other processes since the early days of the solar system are covered in circular scars. They were first seen on the Moon by a shocked Galileo: how could the perfect celestial sphere be so disfigured?

It was thought for centuries that lunar craters were volcanic in origin. The same was true of even the most obvious and pristine impact craters on Earth, including Meteor Crater (as it is now known) near Flagstaff in Arizona. The impact theory was resisted because it was a process that geologists never witnessed in action. Impacts also implied sudden, random catastrophes, something that jarred with the Uniformitarian philosophy of slow, gradual changes, which prevailed

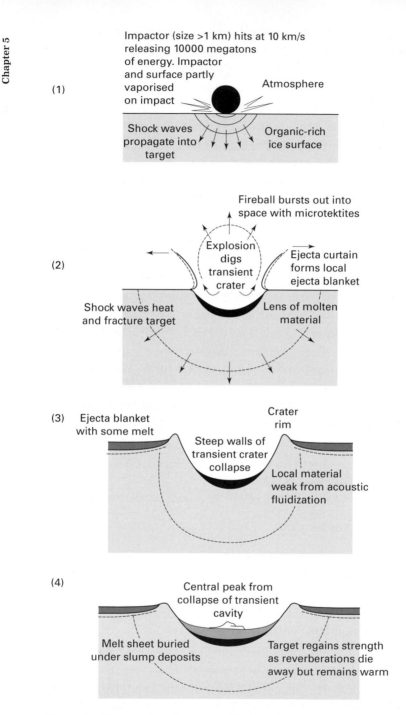

Figure 5.1. The formation of a crater on Titan by a large impactor (1 km or more in diameter). Reid Thompson and Carl Sagan pointed out in 1992 that such an impact would form a melt sheet, where liquid water could react with the abundant nitriles on Titan's surface, synthesising important pre-biotic molecules such as amino acids.

for many years. The impact origin of craters became more widely accepted as similar structures produced by man-made projectiles and explosives became commonplace in the middle of the twentieth century. To date, around 150 natural impact craters have been recognised on Earth.

Impact craters on satellites in the outer solar system are broadly similar in appearance to craters on the Moon, Earth, or indeed, Mars, Venus and Mercury, although there are some subtle differences. The size and shape of a crater is determined by a combination of the energy carried by the impactor (determined by its mass and speed) and the density, strength and gravity of the target.

The smallest craters are described as 'simple'. They are bowl-shaped cavities, usually with slightly raised rims. Above a certain size, called the 'transition diameter' by cratering aficionados, the simple geometry gives way to something with more structure and the descriptive but rather unimaginative term 'complex crater' applies. The rock reverberating with the impact shock can flow like a fluid for a short while. When large craters form this fluid phase lasts long enough for the rock to slosh back into the bowl. The circumference often slumps, sometimes giving rise to a 'terraced' appearance, with flat shelves below the rim. This slumping, and more particularly the rebound of the compressed crater floor, leads to a fairly flat bottom (compared with the bowl-shaped depressions of simple craters) and, most impressively, a central peak. You can get a good idea of how a medium-sized crater forms by watching drops falling into a glass of milk. This model exaggerates the dynamics somewhat but overall it is not a bad demonstration.

In larger craters there is often a raised ring in the middle of the crater rather than a single central peak. Also, in softer material, like ice, the flat floor of the crater can start to bow upwards, dome-like. The largest impact structures tend to have very flat floors. Sometimes walls and floor merge so that all topographic evidence of the original crater disappears leaving only a circular patch of a different colour called a palimpsest. This happens because the cavity excavated by the impact is so deep that the soft crustal material below the surface wells up to fill it. The flow of material at depth drags the crust surrounding the crater along with it. This can crack the crust, which shifts and tilts, producing concentric circular scarps. Structures formed in this way earn the name 'multi-ring' craters. The most spectacular examples are on the Moon and on Jupiter's moon Callisto.

What would all these different kinds of craters look like on a world with lots of liquid? In fact, many craters on Earth are in regions that are wet – Canada for example – so crater lakes are not unknown. Examples include the wide ring of Manicouagan (pictured in Figure 5.3) and the close pair of Clearwater Lakes in Canada, Bosumtwi in Ghana and El'gygytgyn in Siberia. A number of particularly spectacular lakes are in volcanic rather than impact craters, such as 'Crater Lake' in Oregon in the western USA. There are several other examples in Hawaii and New Zealand. Some features on Titan may well look like them, even though not formed in quite the same way.

But how would liquids get in? If the crater floor were sealed, maybe it could fill with rainfall. However, although an impact event can create a pool of molten rock (or molten ice, in Titan's case), which we might expect to seal the bottom of the crater, in general the shock waves and violent ground movements will break up the bedrock, rendering it porous.

So crater lakes might not hold water, so to speak. However, if the ground were soaked, so that the bottom of the crater is below the 'liquid table', then the crater will fill up to that level. This is something that *Cassini* may be able to tell us with its radar altimeter: if there are crater lakes – and we bet there are hundreds – are they all filled to the same level? If so, then they are probably connected by liquid saturating the ground.

Some crater forms may give rise to spectacular landscapes. Imagine

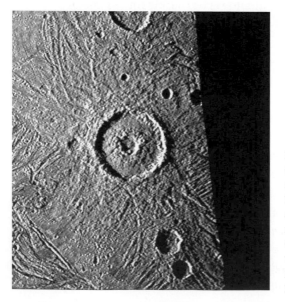

Figure 5.2. Crater Ashima on Ganymede imaged by the *Voyager 2* spacecraft. This 87-km-wide crater has a central pit 23 km in diameter and a visibly domed floor, due to the relaxation of the soft ice interior of Ganymede. Titan craters may look very similar and could perhaps be partly submerged. NASA image processed by Paul Shenk.

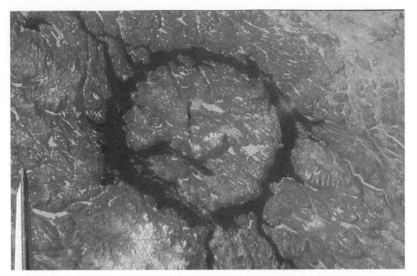

Figure 5.3. The Manicouagan impact structure in Canada is 70 km across. Although the crater itself has long been eroded away by ice sheets, the ring-shaped lake highlights the subtle vestige of the crater. This image was taken in 1983 from the Space Shuttle *Columbia*, whose tail fin is visible in the picture. NASA image.

a multiple-ring crater filled with dark hydrocarbons, looking like a giant bull's eye. Smaller versions of the same thing may be apparent in medium-sized craters with domed interiors and central pits, as described in the science fiction story we mentioned in Chapter 1.

What about impacts into crater lakes? An impact would cause a small crater, although the few examples we have on Earth of submarine craters ('hydroblemes') are rather subdued. The event itself would be spectacular. A huge tsunami would spread out from the impact site, ploughing into and splashing up the crater rim, perhaps shifting boulders that could not otherwise move. The tsunami would be strongest on the side of the crater closest to the impact. Perhaps by mapping out the movement of boulders we could infer where a submarine crater is lurking. Still more intriguing is the idea that an impact could occur in a ring-shaped lake. The circular tsunami would spread out from the point of impact, then as it met the edges of the ring, two huge waves would race around the circumference of the ring, until they clashed together on the other side, jetting upwards and sideways.

In general, when a crater forms, its surrounding collar of ejecta will be confined to a layer of scattered rocks, solidified melt and debris no wider than a few times the size of crater itself. There will be no long rays like those we see around certain craters on the Moon because

Figure 5.4. An impression of a landscape on Titan by space artist James Garry. Rounded hills loom through low fog near a shallow impact basin. Just inside the crater rim, liquid deposits have dried up, leaving a few tarry pools and a rippled lake bed. The crater floor towards the right has domed upwards, while subsurface heat sources remaining from the impact drive hydrothermal-like activity, jetting methane steam from fumarole vents. (In colour as Plate 10.)

Titan's thick atmosphere will slow the ejecta down. Some fine material may form streaks downwind of the crater. We see such things on Mars.

Guessing from the cratered surfaces of the other saturnian satellites pictured by the *Voyagers*, Titan should have about 200 craters 20 km across in every million square kilometres or so, or around 16 000 in all. Larger craters are rarer. Going up in size by a factor of ten leads to a drop in the expected number of craters by a factor of several hundred. Equally, we expect very few craters smaller than about 10 km in diameter because the objects that might excavate such small craters burn up in the atmosphere before they reach the ground.

Craters may tell us about Titan's rotation history. If Titan has always been tidally locked with one face permanently turned towards Saturn, then the leading or 'front' face will hit more interplanetary debris, giving rise to many more craters on that side, which incidentally is the 'bright' side. On the other hand, if the irregular moon Hyperion is just the largest splinter of a what used to be a larger moon before it was broken up by an impact, most of the debris from that event would have rained down on the trailing side of Titan.

Volcanoes of ice

Any volcanoes on Titan will be made of water, or more likely a blend of water and ammonia, since this antifreeze mixture remains liquid to lower temperatures than pure water. The curious form of volcanic activity involving icy material rather than the eruptions of molten silicate rock more familiar on Earth is called 'cryovolcanism'. In the same vein, molten ice below the surface is 'cryomagma', becoming 'cryolava' when it erupts. If the cryomagma comes from a depth of a few tens of kilometres, welling up in a crack through the ice crust, most will make it to the surface, with only a little freezing on the sides of the crack. It will then ooze out over the surface.

There are in fact ice mountains on Earth – about 10 feet high – that may be prototypes of ice volcanoes. Called the Lake Superior Icefoots, they form around the shores of the Great Lakes in North America when the lakes freeze in winter. Wind-blown waves pile into and under the ice sheet. When the ice cracks, often quite far from the edge of the ice, water sprays up through the crack and freezes as it falls back. Soon a conical hill is built up, sometimes spraying water from its top.

Rock volcanoes take a variety of forms, depending on the properties of the lava, in particular its gas content, which determines how fizzy it is, and its viscosity – that is, whether it is runny or stiff. A gassy, viscous lava will build up pressure in the lava duct, then burst out explosively. If it doesn't blow out the side of the mountain, as happened when Mount St Helens in the USA erupted in1980, it shoots dust and ash high into the air. Some of it falls back to create a pretty 'stratovolcano', like Mount Fuji in Japan, with a characteristic conical shape. A low-viscosity magma, without much gas pressure, may dribble out quite gently. This type, common in Hawaii, flows over the surface and congeals, eventually building up a wide, shallow-sloped shield volcano.

If the lava is moderately viscous but does not cool down very rapidly, a pancake-shaped dome can form. The best examples of these are on Venus. US scientist Jeff Kargel, while a student at the University of Arizona, performed some experiments with ammonia solutions, chilling them down to low temperatures and measuring their viscosity. These simple experiments suggested that ammonia-spiked cryolavas would be quite viscous – rather more so than water without ammonia. Based on these experiments, and taking into account the low gravity, the pancake dome shape seems perhaps the most likely for volcanoes on Titan.

The gas content of a cryomagma, which is critical in predicting its behaviour, is hard to estimate. The possibility of a cryomagma column encountering a pocket of methane resulting in an explosive eruption can't be ruled out but, in general, water–ammonia cryomagmas shouldn't be able to dissolve more than a tenth of a per cent or so of methane by weight, according to the very limited data available. At Titan's significant atmospheric pressure this amount of gas is not enough to produce bubbles big enough to break up the magma column to form a stuttering, violent eruption (called 'Strombolian' after the characteristic intermittent eruptions of Stromboli in Italy), so this seems improbable.

Of course, as on Earth, there is no reason to expect all volcanoes to erupt on land. Many may erupt beneath lakes and seas. Dramatic underwater pictures from Hawaii have shown this happening. The

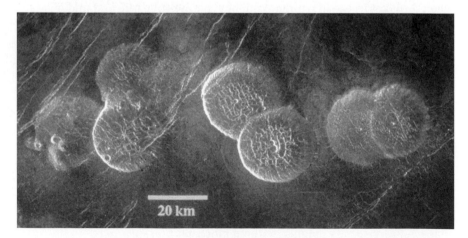

Figure 5.5. A radar image from the *Magellan* spacecraft showing 'pancake' domes on Venus. These volcanoes are formed by the eruption of very viscous lava. They are about 800 m high and 25 km across. It is possible that cryovolcanoes on Titan could have a similar structure. NASA image.

Plate 1 The launch of the *Cassini–Huygens* mission on 15 October 1997 at 4.43 a.m. EDT, from Cape Canaveral Air Station in Florida. The launch vehicle was a Titan IVB/Centaur. NASA image. (*See Chapter 1.*)

Plate 2 A montage of the seven largest planetary satellites in the solar system (*see Table 1.1*). From left to right, top row: the Galilean moons of Jupiter, Ganymede, Callisto, Io, Europa; bottom row: the Moon, Titan, Triton. Images NASA. (*See Chapter 1.*)

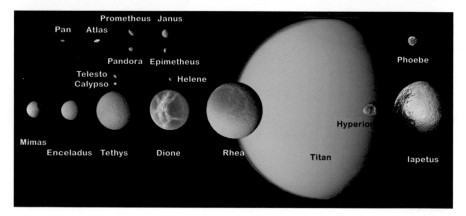

Plate 3 A montage of *Voyager* images of saturnian moons. NASA images. Adapted from a montage prepared by David Seal. (*See Chapter 1.*)

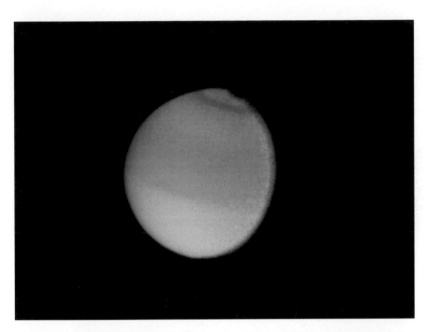

Plate 4 An image of Titan taken by *Voyager 2* on 23 August 1981 from a distance of 2.3 million km (1.4 million miles). The contrast has been enhanced to reveal a lighter area in the southern hemisphere and a dark collar around the north pole. NASA image. (*See Chapter 2.*)

Plate 5 Detached layers of haze in Titan's atmosphere imaged by *Voyager 1* on 12 November 1980 from a distance of 22 000 km (13 700 miles). NASA image. (*See Chapter 2.*)

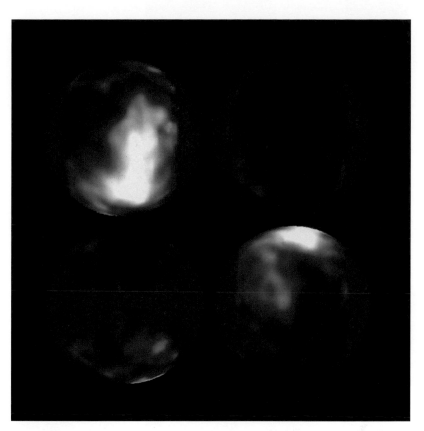

Plate 6 Four views of a global projection of the map of Titan's surface assembled from 14 images taken with the Hubble Space Telescope through the F850LP filter (*see Fig. 2.8*) between 4 and 18 October 1994. The upper left image is of the Saturn-facing hemisphere. Each subsequent image (working from upper left to lower right) represents a rotation of 90°. Thus, the upper right image is the 'leading' hemisphere, the lower left is the 'anti-Saturn' hemisphere and the lower right is the 'trailing' hemisphere. The resolution is about 580 km (360 miles).

The solid brick-red colour around the poles shows areas that could not be imaged through the haze. The gap in coverage extends to the equator at about 10° longitude (top left image). The shading here shows the average intensity. The overall contrast amounts to only about 10% of the total light collected through the filter.

Credit: Peter H. Smith and NASA. (*See Chapter 2.*)

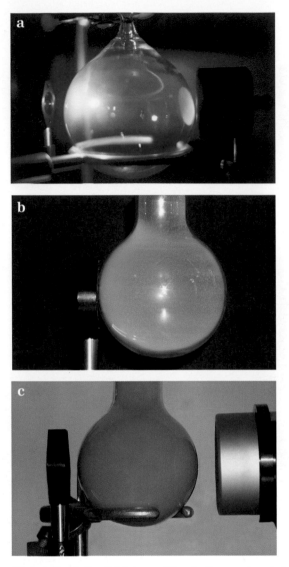

Plate 7 Simulating lightning on Titan. In the experiment pictured here, which was carried out in 2001, a focused (invisible) laser beam was used to make a white-hot plasma in a flask of methane and nitrogen (a). After a short time, the flask became fogged with a white deposit (b). The deposit became brownish as it thickened (c). The brown deposit of organic chemicals resembles Titan's haze. Laser plasma experiments are more convenient than the electrical discharge and ultraviolet illumination experiments conducted by Miller, Sagan and others because the deposit is generated in only a few hours rather than days. Images courtesy of Rafael Navarro-González, Laboratory of Plasma Chemistry and Planetary Studies, National University of Mexico. (*See Chapter 3.*)

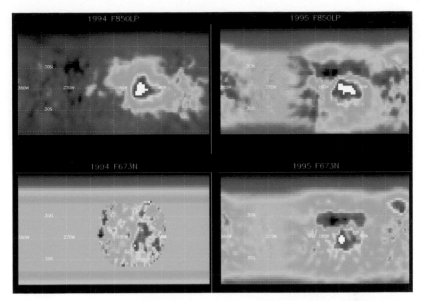

Plate 8 A comparison of maps of Titan made from HST images taken in 1994 and 1995 by a group including Mark Lemmon, Ralph Lorenz, Peter Smith and John Caldwell. Images are through two different filters – the F850LP centred around 940 nm (*see Fig. 2.8*) and the F673N at 673 nm. The large bright feature is apparent in the data for both years. In 1994, the F673N data was only taken around this bright region. The F850LP images primarily sense surface features, while the F673N images barely detect the surface and are quite sensitive to clouds. A bright spot can be seen near the top right of the 1995 maps but it is absent from the 1994 data. Its remarkable brightness near the edge of each image, especially at 673 nm, suggests that this transient feature was a cloud. Image by Ralph Lorenz. (*See Chapter 4.*)

Plate 9 A Hubble Space Telescope image of Saturn taken on 6 August 1995, when the rings were practically edge-on as seen from Earth. The false-colour image was assembled from separate exposures taken through four filters at 336, 467, 619 and 673 nm. The black line bisecting Saturn is the shadow of the rings. The prominent dark circle on Saturn is Titan's shadow. The four moons clustered around the ring plane on the right are, from left to right: Mimas, Tethys, Janus and Enceladus. Credit: Erich Karkoschka and NASA. (*See Chapter 4.*)

Plate 10 An impression of a landscape on Titan by space artist James Garry. Rounded hills loom through low fog near a shallow impact basin. Just inside the crater rim, liquid deposits have dried up, leaving a few tarry pools and a rippled lake bed. The crater floor towards the right has domed upwards, while subsurface heat sources remaining from the impact drive hydrothermal-like activity, jetting methane steam from fumarole vents. (*See Chapter 5.*)

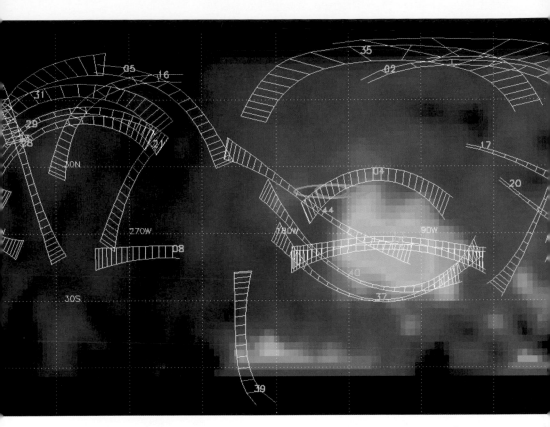

Plate 11 *Cassini's* radar will map all of Titan crudely but will image only about 25% of Titan's surface at its highest resolution. This high resolution imaging will take place during close approaches in the course of some of *Cassini's* many passes by Titan over its four-year mission. Deciding which passes is a matter for negotiation with other science teams. This diagram shows the tracks that might be covered under one proposal. The white radar tracks are overlain on a version of the HST map and are labelled with the Titan flyby numbers (yellow). Their odd shapes are due to a combination of the flyby trajectories and the projection of the map. Some tracks are only half as long as others. On these close approaches, *Cassini* would spend time pointing in another direction to perform a different observation. The areas where the *Huygens* probe is expected to land, with two different assumptions about the wind, are indicated by yellow and green ellipses. Image by Ralph Lorenz. (*See Chapter 6.*)

Plate 12 Technicians at Daimler–Benz Aerospace (as it was then known) begin final assembly of the *Huygens* probe. The back of the heat shield is visible, as is the top of the descent module. The back cover, on a separate wheeled dolly, is towards the left. ESA image. (*See Chapter 6.*)

Plate 13 An impression by ESA artist David Ducros of the *Huygens* probe shortly after release from the *Cassini* spacecraft. When the event actually takes place, Titan will be farther away than depicted here. The artist has also been unable to resist the temptation to show the rings ajar; in reality, they would appear edge-on. ESA artwork. (*See Chapter 6.*)

Plate 14 *(below)* An artist's impression of three possible scenarios for the landing of the *Huygens* probe on Titan. From left to right: the 'crunch' (on solid terrain), the 'squelch' (in slush), and the 'splash' (in liquid). Artwork by James Garry. (*See Chapter 6.*)

Plate 15 Flight mechanics from JPL lower the *Cassini* spacecraft onto its launch vehicle adapter. The adapter was later mated to the Titan IV/Centaur launch vehicle. NASA image. (*See Chapter 6.*)

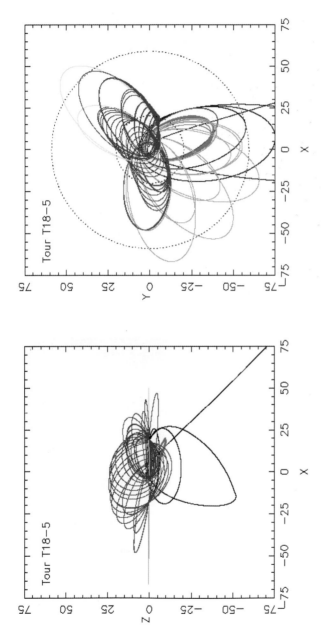

Plate 16 *Cassini's* four-year tour around Saturn (before modifications at the beginning were introduced by the change of plan in 2001). (a) Looking down over Saturn's pole with the Sun off to the right. The axes are labelled in units of Saturn radii. The inner dashed circle marks Titan's orbit at 20 Saturn radii and the outer dashed circle is the orbit of Iapetus. The black section is arrival, Saturn orbit insertion (SOI) and *Huygens* probe release; violet is the early-period reduction phase until April 2005; orange is the occultation sequence ending in September 2005; green is the 'petal rotation' and magnetotail sequence; blue is the July 2006 to 2007 transfer back to the dayside; yellow is the rotation and icy satellite sequence in summer 2007; red is the high inclination sequence at the end of the tour. (b) Looking into the ring plane. The colour codes are the same as in (a). The kink in the black section is the periapsis raise manoeuvre prior to probe delivery. Note the blue 'cranking over the top' sequence and the many small inclined orbits at the end of the tour, shown in red. The rotation sequences coloured green and yellow in (a) are in the ring plane. (*See Chapter 6*.)

Plate 17 One of *Cassini's* most dramatic moments – Saturn orbit insertion in July 2004 – as depicted by JPL artist David Seal. One of *Cassini's* main engines will burn for some 90 minutes as the spacecraft whips over the ring plane. Artwork: NASA/JPL. (*See Chapter 7.*)

Plate 18 (*opposite*) A dramatic Titan landscape as visualised by artist Mark Garlick. The scene includes circular lakes. The lightning and the looming presence of Saturn in the sky may both be rather more subdued on the real Titan. © Mark Garlick, www.space-art.co.uk. (*See Chapter 7.*)

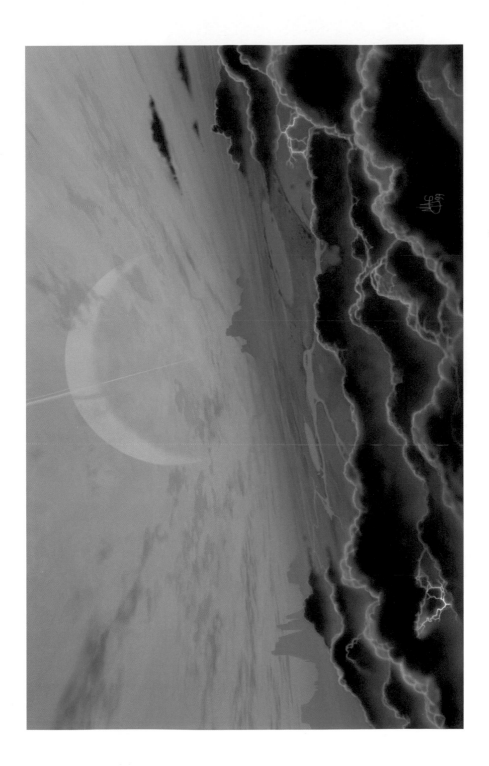

Plate 19 An impression of a post-*Cassini* mission to Titan realised by space artist James Garry in collaboration with Ralph Lorenz. Compared with a 'rover', a helicopter would offer superior mobility and flexibility while taking advantage of Titan's low gravity and thick atmosphere. The design shown here has contra-rotating rotors for compactness and spindly legs to keep the rotors well clear of the ground when it lands. It is shown inspecting the remains of the *Huygens* probe and using spectrometers to determine the amount of haze deposited on the probe since its landing. (*See Chapter 7.*)

lava cools very quickly, oozing out like toothpaste, while water boils around it. It forms rounded blocks, called pillow lava, which stack up in a mound partly welded together. Ammonia-rich lavas on Titan, at the bottom of ethane lakes, might form in a similar way.

No glaciers of sludge

A glacier, though huge and apparently motionless, is very much a dynamic entity – literally a river of ice. Ice accumulates at the high end and is balanced by loss – through evaporation, melting, or calving of icebergs – at the low end. The greater the accumulation rate, whether because of heavier snowfall or a larger catchment area, the faster the glacier flows to take ice away from the accumulation region. Some glaciers, like those in cold and dry Antarctica and Greenland, are slow-moving because loss and accumulation rates are quite low. Some, like those on the west coast of New Zealand, sprint through temperate rainforest where loss rates are enormous but they are balanced by the huge snowfalls (equivalent to 10 m of water a year) on the mountains near the coast. In the past century or so, many glaciers around the world have retreated. They are still flowing almost as fast as ever but, because of the warmer climate, loss rates have increased and the 'break-even' position of glaciers has tended to move uphill.

On Titan we could conceive of glaciers made of organic sludge driven by the incessant drizzle of haze, or glaciers running on methane snow. The problem with a sludge glacier on Titan is that accumulation is rather low. To have a flow, say 100 m wide and 10 m deep moving at a metre a year requires about 1000 tonnes of material every year. A collection area ('cirque') 1 km on a side would have to catch 1 kg of material in every square metre – or a depth of about 1 mm a year. This is orders of magnitude larger than the amount of organic material drizzling down as haze, so that just doesn't work.

Although methane is a good analogue for water on Earth, methane glaciers look unlikely because methane doesn't seem to be solid at Titan's surface temperatures. Although the freezing point of pure methane is about 90 K, roughly the same as the surface temperature of Titan's high latitudes, the antifreeze effect of nitrogen lowers the freezing point below 80 K; lower than the temperatures we believe occur anywhere on Titan – even at the summits of its highest mountains.

Perhaps, though, if Titan had a cooler episode in its past, there could have been methane ice on the surface. But if glaciers were possible under such conditions, it doesn't mean they were a powerful erosive force. U-shaped valleys on Earth are carved most effectively by thick fast-moving glaciers. Ice itself doesn't erode anything except very soft rock but boulders gripped in the ice act as tools or teeth to cut the rock beneath. The low gravity and comparatively small accumulation rates on Titan mean that both the force on the teeth (if there are teeth) and the speed with which they are dragged across the rock would be very small. So the U-shaped valleys that glaciers carve and the lakes or inlets (fjords) that form in them seem unlikely.

Improbable river valleys

River valleys are one of the most common features on Earth's surface. As well as the rivers we see every day, old valleys can even be detected by depth-sounders on the bed of the Mediterranean, carved before the straits of Gibraltar burst and filled the desert basin with water. And the penetrating vision of space-borne radar has revealed ancient riverbeds beneath the sands of the Sahara. However, river valleys seem unlikely on Titan at first sight. Overall rainfall rates are low – there just isn't enough sunlight to drive the methane evaporation to feed many rain clouds. It is tempting to dismiss rivers on Titan.

On the other hand, as we discussed earlier, this 'average view' may be misleading, as it would be in some deserts on Earth – especially if the rain falls in huge downpours that would temporarily increase the local humidity over a wide area and so retard the evaporation of any nascent rivers. Arthur C. Clarke coined the phrase 'methane monsoon' in his novel *Imperial Earth*. And although rainfall should be much less erosive on Titan than on Earth because the drops fall much more slowly in the weak gravity, if the drops *can* conspire to form rivers their flow might be quite powerful, since the viscosity of methane is rather lower than that of water, encouraging turbulence.

The fact that any rivers would be close to their boiling point, cold as that is, could enhance their erosive capability. A similar effect damages ships' propellers. As ships became faster, their propellers became pitted, even though water doesn't usually erode metal. It turns out that a propeller can generate such suction that the pressure close to the blades falls to the point where there is effectively local boiling and bubbles of vapour form, a process called 'cavitation'. It is the bubbles that caused the erosion. Since only a small drop in pressure would be required to cause a methane river to 'boil', it may cavitate as it flows, making it erosive. If rivers carry much debris, either eroded ice particles in suspension or organic sludge, they may perhaps form large deltas where they empty into lakes. We'll see.

The karst that never was

At one time the idea came about that Titan might have a lot of karst terrain. Karst is a peculiar landscape (named after a region in Yugoslavia) that on Earth results from mildly acid rain and ground-water dissolving limestone rock. Typically there are many under-ground cave systems, some of which breach the surface at sinkholes. It is in just such a sinkhole that the Arecibo radio telescope is con-structed. Peculiar rock-shapes like arches can also result and the idea motivated artist Kim Poor's impression of the *Huygens* probe making a dramatic splashdown into a methane lake with a huge natural arch in the foreground.

The karst idea adapted to Titan – basically with methane assumed to dissolve ice – was theory at its best. It solved several problems in one fell swoop. The caves would give liquid methane a place to hide. They could resupply the atmosphere with methane without requiring large exposed areas of liquids on the surface, which were ruled out at the time by the (erroneously high) radar reflectivity. The idea was neatly self-consistent in that liquid methane would create the cave systems in which it could hide. It also conjured imaginative mental images of what Titan might look like. There was only one problem. The whole picture was built on a false premise.

Someone once called it the great tragedy of science, when a beauti-ful theory is destroyed by an ugly fact. The concept of karst terrain on Titan owed its existence to a single experimental indication that liquid nitrogen and liquid hydrocarbons can dissolve water ice. This one experiment had never been repeated and was somewhat at odds with physical theories of dissolution, which predicted that ice would be quite insoluble.

It turned out that the conclusions drawn from the experiment were wrong. The experimenters had overestimated the amount of stuff that dissolved in their test because of contamination in the solution cell. An innocent mistake but a reminder to heed the advice of TV's *X-Files*: 'Trust no-one'.

RALPH'S LOG. AUTUMN 1991.

When I started my Ph.D. in Canterbury, I was in touch with Jonathan Lunine, who having always been uneasy with the single experimental result, suggested I check it out. His enquiries to the first author of the paper had never met with a reply. Since the measurement had been made at Southampton University where I had been an undergraduate, I was able to find out that the first author had been a Ph.D. student and was no longer there, hence the failure to reply. One of the other authors revealed that the spectral absorption they thought was water dissolved in the solution was in fact CO_2 that had crept in, and that they had since published a retraction.

The sands of Titan?

The physics underlying the formation of sand dunes and related phenomena was first articulated by a gentleman named Ralph Bagnold. In the 1930s, he made some of the first expeditions into the Egyptian desert by car (Model T Fords, no less) and became intimately familiar with the desert. Gifted with an enquiring mind and the opportunity, he decided to investigate desert phenomena.

He learned by basic observation that the sand does not float suspended in the air but rather moves along the ground in short jumps. This process is called saltation, derived from the Latin word 'saltus' meaning 'a leap'. At the right time in a sandy desert, you can see the ghostly shimmer of sand moving in a layer a few centimetres thick. Snow is blown in the same way and can be seen writhing snakelike across a dark road, again in a saltating layer

Tear open a bag of granular sugar, for example, and shake it out. All the grains fall to the ground in a second or two. Do the same with icing (confectioners') sugar and you get a dust cloud that lasts some time. The smaller particles fall more slowly and are easier for the air to move because their surface-area-to-mass ratio is higher. On the other hand, small particles stick together better. You can cut a little vertical wall in icing sugar and it will stay put but such a wall in granular sugar will instantly collapse. So it turns out that there is an optimum size for saltation. Particles of this ideal size require the lowest wind speed to get them moving.

There are some equations for estimating this optimum size, based on what the sticking mechanism is assumed to be. If the mechanism is the same on Titan as on Earth, then as one might expect given the lower gravity and the denser atmosphere, the air on Titan can more easily carry particles and the optimum size is larger – about 0.3 mm compared with closer to 0.1 mm on Earth. Also, the minimum wind speed required to lift these particles is considerably less – about ten times smaller than for Earth.

However, all this means very little if the stickiness of particles is different, as well it might be on a moon covered in organic sludge. Electrostatic effects may play a large role too since Titan has no water to dissipate charge build-up. Electric charges generated by friction can make particles clump together or attract them to a surface, like dust on a TV screen. Electrostatic sprayers and electrostatic dust extractors rely on this effect.

Then of course there is the question of how you make sand. On Earth the erosive action of rivers is one major source of sand, helped by rocks carried along on the river bed that clash together, grinding each other up. Some fine particles are created by glacial erosion, volcanic eruptions, the effects of freeze-thaw cycles or wind erosion on larger bodies of rock. All of these processes are likely to be weaker on Titan if they exist at all, so there might not be much ice 'sand' to begin with.

The same arguments apply to Venus, which has, as far as we can tell, a fairly bare surface and certainly no rivers or glaciers. But there is another source of sand-sized particles: impacts. The microtektites thrown out of an impact fireball are essentially sand-sized spheres. In large impacts they can be thrown out of the atmosphere and over large distances. Some impact debris from the dinosaur-killing Chicxulub impact was found 10 000 km away in the middle of the Pacific Ocean. On Venus, deposits of this kind of material form dark parabolic features around the most recent impact craters. The curves all point westwards, since the debris is blown by the zonal winds as it falls back through the atmosphere. On Titan this process does not happen as readily because the atmosphere extends so far into space but long streaks of debris, rather than wide parabolas, should be evident downwind of craters. One estimate suggests that the number of particles a millimetre across or smaller could be equivalent to a layer only 10 cm thick over the whole of Titan's globe – not much to go around, especially if 'sand' accumulates in lakes.

On the crest of a wave

Interesting features on Earth's surface aren't confined to the land area. While the sea is basically flat on large spatial scales (which is why you need a powerboat to water-ski, after all), everyone who has ever been either surfing or seasick knows that locally it may be far from flat.

There are basically two kinds of waves on a liquid surface, distinguished by the nature of the physical force trying to pull the liquid back to a level surface. Tiny waves called capillary waves are due to surface tension, the elastic 'skin' of a liquid that tends to pull drops into spheres. This same force stretches the surface of water to allow pond skaters and other insects to walk upon it. Like the elastic skin of a drum, if the surface is disturbed, surface tension tries to pull it back. Capillary waves act only on small scales – the centimetre-sized ripples that form when a gust of wind blows across a pond. On Titan these would be similar, perhaps a factor of two larger.

The more spectacular and perhaps more familiar type are gravity waves. The weight of the fluid drives these waves back towards the natural flat surface of the ocean. All waves ultimately die away due to the viscosity of the liquid, so to see them in nature requires that something disturb the surface. Wind is the most common cause but occasionally a large disturbance like a sub-sea earthquake or a giant impact can cause a huge set of waves, called by their Japanese name 'tsunami' or occasionally (and misleadingly, since tides have nothing to do with them) 'tidal waves'.

In a shallow liquid (where shallow means that the depth is small compared with the wavelength) the speed of a gravity wave depends on the square root of depth times gravity. If the depth is the same, then a wave on Titan travels more slowly than on Earth by a factor of $\sqrt{(9.8/1.35)}$, equal to about 2.5.

But would there be waves at all? The energy to drive waves comes from the wind and Titan is handicapped in two ways. First, it takes a finite stretch of open sea with the wind blowing across it (the 'fetch') for the waves to build up before they settle down to a steady or 'fully developed' sea. Titan is small and we know from the images we have that the lakes and seas are limited to perhaps 600 km across at most. Secondly, the winds near the surface are probably very weak, since there is so little sunlight to drive them and the thick atmosphere doesn't need to move quickly.

So, by and large, Titan's seas should be placid. However, if there are near-surface winds of around 1 m/s, then modest (20-cm) waves could be generated. According to calculations by Nadeem Ghafoor at the University of Kent, the thicker atmosphere means that such waves could become fully developed with fetches of only 30 km or so. It requires winds three times faster do develop such waves on Earth.

So, the prospects for waves are only modest: surfers are unlikely to get excited by 20-cm waves. But at least they are not impossible. Waves significantly affect the radar brightness of the ocean surface. In fact, satellite radar measurements of Earth's oceans are routinely used to measure the wind speed, so *Cassini* may see the effects of waves from orbit. The *Huygens* probe, if it lands in a liquid, will float and bob around, and will be able to measure the waves if it encounters any.

A tide in the affairs of Titan

Impact craters are not the only landscapes on Earth that are due to the direct influence of celestial bodies. Tidal flats are very wide, shallow expanses of sand and mud found on some coastlines, where the regular rise and fall of the tides does more than the battering of waves to shape the scenery at the meeting of land and ocean. Ocean tides on Earth are raised by the gravitational pulls of the Moon and the Sun, the contribution made by the Moon being about three times stronger than the Sun's. Saturn's gravity similarly exerts tidal forces on Titan, though the effects are different and more subtle because Titan always keeps the same face pointed towards Saturn.

On an idealised Earth as calculated by Isaac Newton, there would be an 18-cm bulge in the ocean, on the side directly facing the Moon and on the diametrically opposite side. In reality, the rapid rotation of Earth and the effects of resonances in basins like the North Sea or the Bay of Fundy tend to amplify the tides somewhat and cause some places to have unusually strong or weak tides, or to miss out on alternate tides.

Calculations make the tide induced on Titan by Saturn 400 times larger than the lunar tide on Earth. This means that the height of the 'equilibrium' tide is around 100 m. However, this is not as exciting as it sounds. First, because Titan rotates so as to keep the same face to Saturn, it is as if the tide is frozen in place. It is always high tide at lat. 0°, long. 0° and lat. 0°, long. 180° and always low tide at the poles and at the centres of the leading and trailing sides. Secondly, Titan's

Tidal forces

Tidal forces in moons and planets orbiting around each other arise because they are bodies with a finite size. In simple terms, the gravitational force they experience varies from one side to the other. The result is a distorting effect. Liquids are more easily pulled about than solids but solids are still put under stress by the forces trying to pull them apart.

To understand how tidal forces arise, imagine for a moment that Earth is reduced to the size of a golf ball. The distance between our golf-ball Earth and the Moon is set to ensure that the gravitational force between them is precisely that needed keep the Moon in its orbit. But in reality, Earth is 12 756 km across. Think of the golf ball marking its centre while Earth expands to its real size. Part of the real Earth is more than 6000 km nearer the Moon. The opposite side of the world is the same distance farther away from the Moon. In accordance with the laws of planetary motion, these parts of Earth would respectively like the Moon to orbit a bit more quickly or a little slower. Of course, the Moon can't oblige. It can only orbit at one rate – to suit the centre of Earth. The disparity sets up a tension. It is strong enough to distort the form of Earth and, by mutual influence, the shape of the Moon too. But the result is far more noticeable on Earth because water can flow easily whereas rock cannot.

Over time, the effects of tidal forces have caused the Moon to settle so that we always see the same side from Earth. It is for the same reason that Titan always keeps the same face directed towards Saturn.

interior and crust may have deformed somewhat into this same bulged shape, so the depth of liquid does not vary so much from place to place.

However, the subtle complexities of the real Titan mean that there are still interesting tidal phenomena. Specifically, the orbit of Titan is not quite circular but is elliptical with an eccentricity of 0.029. This means that Titan's distance from Saturn varies, and so does the strength of its tides, by about 9%. Were there a global ocean, this means a quite respectable 9-m variation in liquid level. Of course, this is a lot of liquid and it has to come from somewhere. Currents form to bring liquid from elsewhere to make up the bulge and these currents ebb and flow throughout one orbit.

Carl Sagan and Stan Dermott argued that, as on Earth, the friction between the seabed and these currents would dissipate energy. On Earth, this tidal friction transfers momentum from Earth to the Moon, with the result that Earth's rotation slows down and the Moon's orbital distance increases by about 2 cm each year. Sediment beds show that 900 million years ago Earth's rotation was about 30% faster corresponding to a day only 18 h long. In a sense Earth is trying to copy what the Moon already does by pointing the same face towards the Moon, so Earth and Moon will end up like two nervous boxers circling each other face to face.

On Titan, the primary effect we expect from tidal dissipation is the evening out of any difference between its orbital and rotational periods. As Titan always points the same face at Saturn, this has already happened. Most satellites in the solar system, like our Moon, rotate synchronously this way. Those that do not are usually very small or are far from their primary so that tidal forces are weak.

The second effect we anticipate, as pointed out by Sagan and Dermott, is that the eccentricity of Titan's orbit would decrease to a very small value. The amount of damping or dissipation would depend on the depth of the ocean. In a shallow ocean, the currents would have to be faster, and the dissipation higher, in order to allow the tidal bulge to grow and shrink through one Titan day. They calculated that an ocean on Titan would have to be more than 300 m deep for the eccentricity not to have been reduced to zero over the age of the solar system. Even if the starting eccentricity were large and the ocean shallower than 300 m, the eccentricity today would be smaller than the 0.029 we observe. This calculation remains basically correct and was perceived in the early 1990s as an important constraint on Titan's composition. This tidal argument said that dissipation has to be small for the eccentricity to stay as high as it is. For the dissipation to be small, the ocean must either be non-existent, or deep. A deep ocean would cover most or all of the surface and would appear dark to radar, which it turned out not to be: therefore, no ocean. But then, where does the methane come from and what happened to all the ethane that would be produced? Maybe this tidal argument, elegant as it was, needed closer study.

William Sears, a Ph.D. student working with Jonathan Lunine in Arizona, developed a big numerical model to tackle the tides problem. Just like a GCM for climate, this model worked out the forces and dynamics on the ocean and allowed the ocean to respond. He could look at the flow speeds everywhere and study the effects of a large block of land, or an ocean confined only to low latitudes, and so on. His results basically agreed with the simple analytic model by Sagan and Dermott.

RALPH'S LOG. LATE 1993 TO 1994.

Frank Sohl, a Ph.D. student at Kiel in Germany, had modelled tidal dissipation on Jupiter's moon Io. It would be a straightforward matter, if nontrivial, to substitute models of Titan's interior for Io's to see how much tidal dissipation might occur in the interior. I got Sohl interested in the problem and he started to build the Titan model. We also teamed up with William Sears in a trinational transatlantic collaboration of humble students. It was my first real collaborative venture and a lot of fun.

We exchanged almost all of our calculations, comments and drafts of a paper by e-mail. Often Sohl and I would exchange e-mails and pass on a question to Sears, eight time zones away, usually to get an answer the next morning. I remember one evening in the summer of 1994 when there was a partial solar eclipse. In near real-time, I described the appearance of the partially occulted Sun through my window in an e-mail to Sears. Minutes later he described how the Sun had much less of a bite taken out of it than I saw, due to his different viewpoint. Communications make the world small but astronomy reminds us how large it still is.

Together with Sears and Lorenz, Sohl constructed a number of plausible models and found that if Titan had the most likely structure, with the rock separated into a central core, an outer ice crust and a liquid water–ammonia mantle in between, the tidal dissipation in the interior would be enough to cause the eccentricity to decay whether there was an ethane ocean on the surface or not. This was 1994, several months before the HST map appeared.

The following year, inspired by the HST map showing a variegated surface and most definitely not a global ocean, Dermott and Sagan themselves re-examined their work of 13 years before. What if the ocean were not global after all but there were instead isolated lakes and seas, perhaps in crater basins? Confined in isolated lakes, the liquids cannot fully respond to the changing tidal force and so dissipation would be small.

Thus Sagan and Dermott had refuted their own argument of 13 years before and Sohl and his collaborators had shown it was irrelevant anyway. On the one hand, a realistic array of surface seas would not damp the eccentricity by much but, on the other, the interior could do so anyway. The tidal argument was irrelevant to the depth of Titan's seas. Trust no-one.

There is a counterpoint to all this. If the dissipation in the interior were significant, then that still begged the question of how Titan's orbital eccentricity could be so high. Something relatively recent in solar system history must have changed Titan's orbit – increasing its speed by about 80 m/s or 160 miles per hour. An obvious idea is that a large impact could have given Titan a push but that would need an object around 1000 km across, large enough to smash Titan apart. Sohl and collaborators showed that an object 2000 km in diameter could cause the required change in Titan's orbit, without actually hitting Titan, like a gravitational slingshot effect in reverse. The origin of the eccentricity of Titan's orbit may remain a mystery.

Geysers?

At present we have no idea whether methane is continuously seeping from Titan's interior. It may be that methane continues to dribble into the bottom of lakes. A similar process happens in some volcanic lakes on Earth, occasionally with fatal results. Lake Nyos in Cameroon lies in a volcanic area and carbon dioxide gas trickles into its floor continuously. At the bottom of the lake, the high pressure keeps the gas confined. Ten metres depth of water corresponds to 1 atmosphere (1 bar) of pressure, so a bubble released at the bottom of a 10-m lake, where the total pressure is 2 bars, will grow to double its volume as it rises to the surface. Now imagine a parcel of water at the bottom of a lake containing some small bubbles. If it rises, the bubbles grow and its total volume increases. As the volume of the parcel increases, its density drops and the parcel of water becomes more buoyant.

So if there is more than a certain amount of dissolved gas at depth, the lake can be unstable. Once a parcel of water begins to rise, the pressure falls and the bubbles grow, so the parcel becomes more buoyant and rises further and faster. This is what happened at Lake Nyos in 1986. The CO_2 had built up in the bottom waters, which became unstable. The lake waters turned over buoyantly, bringing the dissolved CO_2 out of solution as gas in an eruption. The huge cloud of CO_2 gas released, being denser than air, flowed across the ground and suffocated most inhabitants of a nearby village.

The eruption of geysers operates in a similar way, except that it is the warmth of the water and the dependence of its vapour pressure on temperature that cause the eruption. Imagine a column of water under the ground. At depth, the weight of the water column above means that the pressure is higher than at the surface and so the boiling point is higher. Now imagine that the column of water is being heated at the bottom. Eventually the base of the column will reach boiling point and bubbles of steam will form. If the column is heated very quickly and only at the base, the bulk of the column will remain cold. The little region boiling at the bottom may make a few bubbles and that is all. On the other hand, if the column is heated somewhat all along its length, much of the column may be close to boiling when the bottom begins to bubble. If that is the case, then the slight upward displacement of the column may push some other regions to shallower depths, where the pressure and hence the boiling point is lower and so they may boil too. Thus the onset of boiling at depth can cause the whole column to boil at once.

Could any of these processes operate on Titan? We don't know about the addition of methane at the bottom of lakes. Substantial heating from Titan's interior seems unlikely but can't be ruled out. On average, geothermal heating on Earth isn't enough to cause any kind of instability but, in some areas such as Iceland, New Zealand and Yellowstone National Park in the USA, the geothermal heat flow is locally much higher than the planetary average. If internal heat flow in Titan has similar (fiftyfold) enhancements geysers could occur there too.

Methane–nitrogen liquids have quite strong variations of vapour pressure with temperature, which partly compensate for the fact that the low gravity means the boiling point does not increase sharply with depth. In other words, there is nothing to exclude geysers on Titan in principle. But even if they exist, they might be hard to find. A spacecraft making 45 flybys of Earth at a range of 1000 km and a single parachute-borne lander would be unlikely to spot an erupting geyser on Earth, since they are so rare. So, our prospects for seeing such things on Titan are not great.

A Titanic experience

So what would it be like on Titan for a human explorer? Walking on Titan might be a little like walking on the Moon. Because gravity is lower than on Earth, the natural period of a step (roughly half the swinging period of a pendulum about the length of a leg) would be rather longer than on Earth – something like two or three seconds rather than one second. Just as the astronauts did on the Moon,

Titan's explorers would probably evolve a sort of bounding gait. (In fact the scaling of walking and running speeds is mathematically similar to the scaling of water waves.)

In a way, it would also be like walking at the bottom of a swimming pool. If you blow most of the air out of your lungs you sink but are only a little less dense than water, so it is hard to push much with your feet on the bottom. Also the drag of the dense water slows you down if you do start moving forward. Titan's atmosphere is less dense by a factor of 20 than water but four times denser than air so the extra drag, though weak, is noticeable. A sprint on Titan would be hard. As with skiing, the difficulty would be keeping in contact with the ground to maintain speed. Skiing on Titan would be rather different and perhaps impossible. Because weight on Titan is so much less, there is less force available to overcome the 'static friction', the stickiness that holds you fixed on a surface. With less weight and thicker air, an Olympic downhill skier on Titan could only go about a fifth as fast as on Earth.

Running might be difficult but Titan could just be the place to learn to fly. Birds can fly on Earth because they are light and because they have limbs adapted for throwing down large quantities of air. Pushing air down produces an up-force on the wings that balances their weight. This momentum flux is most obvious with a powerful shower or hose pipe. The flow of liquid pushes back on the shower head or hose nozzle, causing it to writhe around if it is not held carefully. And so it is with birds; they must push down a certain amount of air every second, in order to balance their weight.

Humans can of course glide on Earth but that is cheating in some sense. The only way to fly unassisted has been with very large, light aircraft like the Gossamer Albatross. Only by keeping the airframe as light as possible, and by having a ridiculously large wing that can force a large amount of air downwards slowly, can a feeble human lift him or herself off the ground.

On Titan, it is a different story. There a human's weight is less by a factor of seven and the dense air is easier to push against. With wings a few metres across, a human could flap into the air and fly at least a short distance. A human's arms are not its most powerful muscles, so serious aeronauts would invent a contraption that would exploit their legs somehow – an ornithopter, or pedal-powered aeroplane, perhaps.

Since Titan's lakes may have waves, perhaps one could surf. Lakes are not the most popular surfing locations on Earth, because they are small. This matters because the distance over which the wind can transfer its energy to the liquid is small and so the energy that can build up in the waves is limited. That said, if waves are generated, they are large but slow. Surfing on Titan would be a slow-motion affair, perhaps more like sailing on Earth. Of course, if the winds are strong enough to whip up waves then they are probably strong enough to permit windsurfing or sailing. Perhaps a sailor could use a kite, pulled by the zonal winds at high altitude, to sail eastwards at least.

Swimming on Titan, even assuming one could do so with a moderately thin and flexible heat-suit and breathing apparatus, would be very different from swimming on Earth. The main factor is the density of the liquid. Most likely mixtures of liquid nitrogen, ethane and methane have a density of $600 \, \text{kg/m}^3$, or about 2/3 that of water. People in general have a density only fractionally higher than that of water – it depends whether you take a deep breath or not. But on Titan, people would sink because they'd always be significantly denser than the liquid.

Mass, weight and newtons

In everyday talk we use 'mass' and 'weight' as if they mean the same thing. To a physicist, they are completely different concepts. What we measure in kilograms or pounds is really mass, the quantity of material in an object. A 70-kg person on Earth is still a 70-kg person on Titan. When we say someone 'weighs' 70 kg, we are being a bit lazy, using short-hand for 'he has the weight of a 70 kg mass'. This is all very well until we want to compare weights and forces on different moons and planets. Then we have to remember that true weight is the pull of gravity on a mass and is strictly speaking a force. Forces are measured in newtons, abbreviated to N). One newton is the force of a 1-kg weight accelerated by $1 \, \text{m/s}^2$. Coincidentally, it is also the weight of a typical apple on Earth. So our 70-kg person on Earth weighs $70 \times 9.8 \, \text{N}$, since the acceleration due to Earth's gravity is $9.8 \, \text{m/s}^2$. On Titan he weighs only $70 \times 1.35 \, \text{N}$.

A human being has a mass, including an allowance for the breathing apparatus etc. of about 90 kg. On Earth, with a gravity of 9.8 m/s, this corresponds to a weight of about 900 N. On Titan, he or she would weigh only 120 N. The buoyancy exerted by the liquid is the weight of the volume of liquid displaced by the immersed body – about 900 N on Earth, or 80 N on Titan. Thus the net downward force on an immersed human on Titan is the difference between 120 and 80 N, or about 40 N. Sharks on Earth are in this negatively buoyant situation. If they stop swimming they slowly sink but it is easy to swim fast enough to generate some 'lift' to balance the extra weight. A human on Titan would similarly sink, though with very modest effort could stay afloat. But the combination of a slightly less dense fluid and a much lower gravity opens up a new possibility – porpoising. Dolphins sometimes do this, swimming vigorously to keep up their speed but spending as much time as possible arcing through the air. Although it looks (and is) energetic, it is actually quite an efficient way to travel if you need to break the surface to breathe. The low drag of the leap is enough to compensate for the behavioural complexity of porpoising and the extra drag caused by frequently breaking the surface.

If you kick vigorously and wave your hands around, you can push yourself about half way out of the water. Half out of the water you only have half the buoyancy and since your weight is unchanged, the difference must be balanced by the thrust you are creating, or 400 N or so. Now on Titan, flailing your limbs around will generate less thrust because the fluid is less dense but only 40% less since liquid hydrocarbons are only somewhat less dense than water. But 40% of 400 N is 160 N – more than the weight of a human on Titan.

So to recap, a human on Titan has a weight of about 120 N and, if fully immersed, a buoyancy of about 80 N. If he or she swims hard, an extra 160 N of thrust is available. Now imagine you are a metre below the surface on Titan. If you do nothing, you will slowly sink but instead (remembering this book) you swim for all you're worth; there is 120 N pushing you down but $80 + 160 = 240$ N pushing you up. Ignoring drag (which is fairly small at these low speeds), you will accelerate at $(240 - 120)/90 = 1.3$ m/s^2, or 1/7 g upwards. In a distance of 2 m, you will reach about 2.5 m/s. You would keep moving upwards for about 2 s after you broke the surface (2.5 m/s divided by the 1.35 m/s^2 gravity downwards). If you took off vertically, your feet would be about a metre above the surface before you fell back.

The sporting possibilities are fascinating. A water polo/basketball variant could feature some awesome jumps and slam-dunks. Or what about a swimming/hurdling hybrid – 10-m swimming stretches separated by walls, slightly higher than the 'waterline', above which the racers would have to leap like salmon. The reader is invited to think of more ideas.

RALPH'S LOG. SUMMER 2000.

A TV company making a documentary about the moons of the solar system called me to ask about Titan and I let slip the business about flying. Me and my big mouth. I'd also let slip the porpoising idea and the next thing I knew I was swimming in Mono Lake in California at 6 a.m., in the buoyant but unpleasantly alkaline waters. Later that day I had a pair of wings strapped to my back. Hopefully these antics may inspire someone to think that Titan is an interesting place. At least I got a free pair of flippers and a ride in a hot air balloon out of the deal.

Ball sports would be very different. Of course, most balls are made of elastic materials like rubber, which become rigid and brittle at Titan's low temperatures; a golf ball would shatter when struck on Titan. For now we'll pretend that the balls can be cleverly heated. The thicker atmosphere and low gravity would make most ball sports into

Figure 5.6. An impression by artist Craig Attebery of the surface of Titan with the *Huygens* probe parachuting down. In reality, the rings of Saturn would not be seen nearly as clearly as shown; the rings would be more edge-on and the *Cassini* orbiter would be invisibly small and distant. NASA artwork.

slow-motion events – and harder work if you wanted things to move fast.

And when all the sports are over, what would you see in the sky? Obviously everything is a dim red. There might be some weather going on too. Saturn itself to human eyes would be at best barely visible at night – a bright fuzzy patch in the sky, like a full Moon all but hidden by cloud on Earth. If we could see at slightly longer wavelengths, it would be spectacular, however. Saturn would be essentially as bright as a full Moon on Earth but would be 12 times larger than the Moon as seen from Earth, occupying a full 6° of sky.

Of course, you would only ever see Saturn from one side of Titan. The view of the Sun and Saturn, and perhaps Saturn's moons, might be rather poorer at some times of the year than others, owing to the seasonal cycle of the haze. Saturn would show phases, just like the Moon, and every 15 years, when the Sun passes through the ring plane, there would be a period when Saturn would cast Titan into shadow. During such events, the refraction and scattering of sunlight in Saturn's atmosphere might make a ring all around Saturn, even though Saturn's disk is much larger than the tiny Sun as seen from Titan and would blot the direct view of the Sun completely. An eclipse could last as long as 6 h. During this period, observers on Titan might be able to see auroras on Saturn and perhaps flashes of lightning against the night-time atmosphere.

Dusk on Titan would be slow and gradual. The scattering of light in the atmosphere means there are no long sharp shadows at dusk, just a gradual deepening of the night-time gloom. It is possible that at some wavelengths the sky never completely darkens, even at midnight. Saturn's rings would of course be edge-on and so barely visible, if at all. The other satellites would make attractive ornaments to Saturn at night and especially during an eclipse.

A guessing game

The history of planetary science is replete with predictions, often laughable in retrospect. The vista we have described in this chapter sums up what we know about how planetary landscapes are shaped and how these processes might be similar or different on Titan. Our success will be a measure of how well we understand planets.

Some processes, like plate tectonics, are so poorly understood that we haven't even discussed them in this chapter. The properties of different phases of ice are such that icy satellites like Titan can undergo convulsive contraction or expansion as they evolve and these processes may be responsible for the 'wispy terrain' seen on Dione or the huge rift valley on Tethys known as Ithaca Chasma.

One thing is certain. Some of our guesses, educated as they are, will be wrong. And Titan is bound to have some bizarre landforms that we haven't thought of, which perhaps have no analogues elsewhere in the solar system. Christiaan Huygens realised the immense difficulty of prognosticating when he wrote

> But indeed all the whole story of Comets and Planets, and Production of the World, is founded upon such poor and trifling grounds, that I have often wonder'd how an ingenious man could spend all that pains in making such fancies hang together. For my part, I shall be very well contented, and shall count I have done a great matter, if I can but come to any knowledge of the nature of things, as they now are, never troubling my head about their beginning, or how they were made, knowing that to be out of the reach of human Knowledge, or even Conjecture.

We have not been deterred from trying but the only way to know for sure what is down there is to go and look, which brings us to the other Huygens and our next chapter.

The *Cassini–Huygens* mission

Nearly all our hopes for uncovering Titan's secrets are pinned on a robotic monster, the size and weight of a large dinosaur. With 7.5 miles of wire for a nervous system, it weighs 5.5 tonnes and is 6.8 m long. The multibillion dollar enterprise to get this surrogate explorer to Saturn spans nearly two decades and involves 15 countries. By the end of its mission, around 7100 work-years will have been spent on it.

Genesis of *Cassini*

A mission involving an orbiter around Saturn and one or more atmosphere probes was already being thought about in the late 1970s, just as the *Voyagers* started on their way. It was the logical next step and the planning of the corresponding mission for Jupiter, what was to become *Galileo*, had already started. One of the earliest NASA ideas was 'SO2P' – a Saturn orbiter with two probes, one for Saturn (which would make a useful comparison with the results of the *Galileo* probe into Jupiter's atmosphere) and one for Titan. But realistic plans for a Titan probe couldn't be made until after the results of the *Voyager* flyby established how thick Titan's atmosphere was, information crucial for the design of heat shields and parachutes.

Thinking started in earnest soon after the *Voyager* encounters. The original idea, as proposed by a consortium of European scientists to the European Space Agency ESA, was that NASA would supply a duplicate of its *Galileo* probe and Europe would build an orbiter – the mother ship that would deliver the probe. A joint NASA–ESA study team was set up. The NASA study scientist was JPL's Wes Huntress, who later went on to be the NASA Associate Administrator for Space

Science – the top job in Space Science in the USA; in ESA Jean-Pierre Lebreton, an affable French scientist, led the study.

Soon it was decided to reverse the agency roles: ESA would provide the probe, while NASA would provide the orbiter. The orbiter would be based around a new 'bus', called *Mariner Mark II*, after the successful *Mariner* series of the late 1960s and early 1970s. Spacecraft engineers use the term 'bus' as auto engineers might use the word chassis – a basic structure around which different variants of the same basic vehicle could be easily built, without 'reinventing the wheel'. For a spacecraft, this does not mean only the mechanical structure but also many of the complex systems, like propulsion, power, attitude control (pointing) and so on.

The first two spacecraft in this series, which (as they all do) promised to reduce costs, were to be *Cassini* and *CRAF* (*Comet Rendezvous and Asteroid Flyby*) which, as its name suggests, was to explore some of the smaller bodies of the solar system. One of *CRAF*'s distinctive highlights was that it would fire a penetrator – like a large instrumented harpoon – into the comet to measure some of its properties *in situ*, while the mother ship orbited overhead. Both *CRAF* and *Cassini* would have scan platforms with cameras and spectrometers, and instruments to study the dust, gas and plasma in their respective environments.

New challenges for Europe

In 1984, an assessment study was performed in ESA – a preliminary investigation to identify the principal technical challenges. A planetary probe was a new venture for ESA, whose only interplanetary spacecraft was *Giotto*, at that time on its way to fly past Comet Halley in 1986. The fiery heat of entry into a planet's atmosphere was a new challenge and such things as parachutes were also a novelty for engineers who were more used to solar panels and thruster nozzles.

The entry protection was a delicate issue. It would be crucial to the success of the project but was also tied up with strategic concerns. The sciences of hypersonic aerodynamics and aerothermodynamics are intimately linked – as is rocketry – with the intercontinental delivery of nuclear weapons. Within the ESA member states, and remember that ESA is an agency with an exclusively peaceful mandate, only the industries of Britain and France had experience with such materials, and much of that experience would have to be kept secret.

In the assessment study phase, engineers in ESA and industry developed impressive and imaginative ideas. One study examined spacecraft autonomy. A Titan probe would be too far away for ground control to respond to problems. The study mapped out ways – essentially implemented only later in NASA's 1998 *Deep Space 1* mission – whereby onboard computers could respond to an unknown environment or equipment failures. For example, if a battery failed, reducing the total amount of energy available for the mission, it would shut down some lower-priority experiments to save energy for the most important near-surface studies. Other creative concepts addressed the aerothermodynamics, advocating the use of exotic materials like carbon composites and beryllium, a marvellously light, hard metal with a high melting point but one that is notoriously difficult to work with.

After the assessment study, ESA then went on to fund a 'Phase A' study – still only paper, but with the aim of developing a credible overall design, with a reliable cost estimate. In the absence of commitment from the US, the start of the Phase A study was prudently delayed by a year to November 1987, to synchronise NASA's schedule with ESA's. ESA's space science budget is fairly small and so there are few missions under development at one time. The Titan probe was in competition with other candidates for 'M1', ESA's first 'medium-class' mission, including *Vesta* (a mission to rendezvous with the asteroid of that name) and astronomy satellites. The Titan probe was selected in November 1988 pending NASA approval of *Cassini* and was named *Huygens*.

Meanwhile, in the USA, *Cassini* had bubbled up to the top of NASA's to-do list and the US Congress approved the start of the project in 1989. NASA and ESA jointly solicited proposals for scientific investigations. Americans and Europeans would compete to provide instruments for both the Titan probe and the orbiter. ESA also invited proposals from European industry to build the probe itself, which would cost around 300 million accounting units (an ESA 'currency', approximately equal to one US dollar or euro). Things were now getting serious. In October 1990, the selected experimental payloads for *Huygens* were announced.

In many respects the most important experiment, and certainly the most massive, was the gas chromatograph/mass spectrometer (GCMS). This 18-kg experiment will analyse both the composition of atmospheric gases and the material delivered from the aerosol collector and pyrolyser. Because Titan's atmosphere is so chemically complex, a simple mass spectrometer (as flown on *Galileo*, for example) would not be able to distinguish the many chemicals that might have the same molecular mass. The GCMS identifies compounds not only by molecular mass but also by their affinity to special coatings inside thin tubes, along which different compounds take different times to travel. The combination of the two techniques is very powerful. The GCMS has dozens of valves, high voltage components, delicate filaments and other design challenges. It was to be tackled by a team led by Hasso Niemann of NASA's Goddard Spaceflight Center. Hasso had built the mass spectrometer on *Galileo* as well as a number of similar instruments.

The second US-led experiment was the descent imager/spectral radiometer (DISR). As well as taking pictures looking down and sideways, this instrument will measure the spectra of sunlight filtering through the haze and reflected up from the ground. It will also measure the sunlight scattered around the Sun (the aureole), which will yield data about properties of the haze. The team is led by Marty Tomasko of the University of Arizona. The CCD detector was provided by Uwe Keller of the Max-Planck Institute for Aeronomie in Katlenburg-Lindau in Germany, who built the camera on ESA's *Giotto* spacecraft to Halley's Comet. The infrared spectrometer came from Michel Combes of the Paris Observatory working with other spectroscopic and imaging experts from the USA, France and Germany.

One of the most conceptually simple experiments on the probe is the Doppler wind experiment (DWE). The only hardware associated with it on the probe is an ultrastable oscillator on one of the two channels of the probe radio link. By measuring the change in frequency of the signal received, aided by a second reference oscillator on the orbiter, the Doppler shift due to the probe's motion can be derived. Most of this Doppler shift is the result of the orbiter's rapid approach towards Titan and some is due to the probe's vertical descent. These two components can be calculated and removed. Any remaining component of the probe's motion is due to wind. The DWE team is led by Mike Bird of the Radio Astronomy Institute at the University of Bonn in Germany. German, Italian and American colleagues rounded out his team.

The aerosol collector/pyrolyser (ACP) is a novel instrument that sucks air through a tiny filter held out in front of the probe, trapping aerosol particles. The filter is then pulled inside the instrument and heated in a set of temperature-controlled ovens. The aerosol material breaks down under heating and the gaseous products released are transferred through a pipe to the GCMS instrument for analysis. The development effort is led by Guy Israel of the Service d'Aeronomie in Paris. Guy had built a similar, but simpler, instrument for the Soviet *VEGA* mission to Venus some years before. His team included Austrian electronics providers and collaborators from the GCMS team in the USA. Although conceptually simple, the instrument posed some severe technical challenges and the two filters originally proposed were reduced to only one.

The *Huygens* atmospheric structure instrument (HASI – originally just ASI until the Italian Space Agency was formed with those same initials, requiring 'Huygens' to be added) will measure the basic properties of the atmosphere during the entry and descent. An organisational nightmare, tackled with aplomb by Italian Principal Investigator (PI) Marcello Fulchignoni, this experiment brought together temperature sensors and central electronics from Italy with pressure sensors from Finland, accelerometers assembled in the UK and electrical sensors and electronics built by a consortium with members in ESTEC, Austria and Spain. For good measure, scientific expertise from the USA, Germany, Israel, France and Norway added to the team. There is probably scope for a joke somewhere: 'Heaven is where the Bankers are Swiss, the Chefs from France, the pressure sensors from Finland . . .'.

Lastly (and it always gets mentioned last) is the surface science package (SSP). This is a collection of small sensors (similar to HASI in that respect) led by John Zarnecki, formerly of the University of Kent at Canterbury, now at the Open University in the UK. As well as an accelerometer and penetrometer (see later) to characterise the impact with the surface, the package includes tilt sensors to measure the probe's orientation and bobbing in any waves; the SSP team was unique in including an oceanographer. An ingenious refractometer will measure the ethane/methane composition of an ocean, with additional information provided by thermal, density and electrical permittivity sensors. An acoustic sensor would act as a sonar to measure the depth in a liquid. While most of the equipment was built in the UK, the acoustic sensor was provided by ESTEC and thermal sensors were developed in part in Poland. An X-ray fluorescence sensor, which was to have been provided by the USA, had to be dropped for budgetary reasons.

Although no radar science instrument was selected, a capability for HASI to digest the signal from the probe's engineering altimeter was added later. The option to include a nephelometer, which would measure the light scattering by haze particles using a laser and a small mirror (actually it was the flight spare from the *Galileo* probe), was also studied but determined not to be worth the effort (and the 5 kg) since DISR would measure essentially the same things.

The contributors of experiments on *Huygens* are distributed fairly uniformly across Europe and the USA. It is interesting to note that being the PI for an instrument is more an organisational job than a scientific one. Of the *Huygens* PIs only Tomasko has worked on Titan before.

More tools for Titan

Huygens is dedicated to probing Titan but there are also several experiments on the *Cassini* orbiter that will be used to study Titan.

Cassini carries long, sensitive radio antennae to search for electrical phenomena in the saturnian system and will listen out for lightning at Titan. A similar search by *Voyager 1* didn't reveal anything but the flyby was over quickly. *Cassini* will be able to detect much weaker discharges in Titan's atmosphere, if they do occur.

A glorified compass-needle sounds an unlikely tool with which to detect an ocean hundreds of miles beneath the icy surface of a satellite and yet that is exactly what the *Galileo* spacecraft's magnetometer did. *Cassini* also carries a set of sensitive magnetometers. Unlike ice and rock, water conducts electricity, especially if it is salty. A conductor moving through a magnetic field will generate a small magnetic field of its own – and a conducting ocean orbiting through a giant planet's magnetic field does just that. While this was no surprise for Europa, dull, dead Callisto showed the same magnetic signature, prompting one distinguished scientist a few weeks later to do calculations, which, he claimed, showed that 'of course' Callisto should have such an ocean. In the case of Titan, a conducting liquid layer seems quite likely. The icy Galilean satellites of Jupiter including improbable Callisto, roughly the same size as Titan, seem to have them. What's more, the ammonia that may have been incorporated into Titan acts as an antifreeze, prolonging the existence of a liquid layer that might otherwise freeze.

Cassini has a radio communications and tracking system more

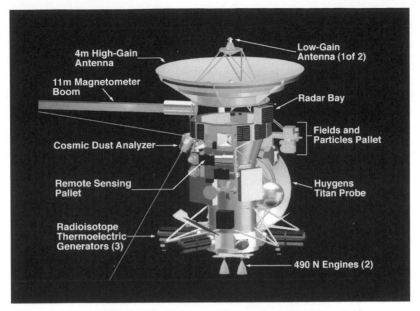

4m High-Gain Antenna

Low-Gain Antenna (1of 2)

11m Magnetometer Boom

Radar Bay

Cosmic Dust Analyzer

Fields and Particles Pallet

Remote Sensing Pallet

Huygens Titan Probe

Radioisotope Thermoelectric Generators (3)

490 N Engines (2)

Figure 6.1. Diagram of the *Cassini* spacecraft in its cruise configuration. Various components and instruments are labelled.

elaborate than any other planetary spacecraft. Passing the radio signal through the upper atmosphere of Saturn or Titan can determine the density of free electrons. Lower down, the radio signal is refracted by the denser layers of the atmosphere (whether Saturn's or Titan's). The radio signals, sent on three different frequencies simultaneously, are also affected by Earth's ionosphere and the tenuous gas between the planets. However, by measuring the three frequencies simultaneously, these effects can be removed to get a much more accurate measurement.

Around Saturn, the radio system can be used to measure the tiny changes to *Cassini*'s trajectory introduced by the gravity fields of its satellites. This in turn can be used to determine the masses of the satellites (although masses are already known for several of them from analysis of their effects on each other and on the rings) and more interestingly, the distribution of mass within each satellite.

Another versatile instrument is the visual and infrared mapping spectrometer (VIMS). Its field of view is a narrow strip but each strip is a line of pixels, each analysed as a spectrum with powerful wavelength resolution. By sweeping the strip around the sky, or simply letting spacecraft motion drag the strip across a satellite, an image cube can be built up – a stack of images of the same scene in many

177

hundreds of different colours. Subtle changes in the brightness of surfaces in specific colours are signatures of particular materials – methane or water ice, for example.

For the first time among all outer solar system spacecraft, *Cassini* has a radar. In fact it was one of the original pre-*Voyager* ideas, that if Titan's atmosphere were opaque, then a radar would be needed to see its surface, as was the case for Venus. *Cassini*'s radar has to be very versatile. Unlike most radar mapping spacecraft, which have a fairly limited range of altitudes of operation, *Cassini*'s will operate from perhaps tens of thousands of kilometres away, beginning an hour or so before closest approach and sweeping its spot-shaped radar in a spiral pattern to measure the temperature and radar reflectivity of broad regions of the surface. Then in the 20 minutes or so around closest approach, when *Cassini* is within 4000 km of the surface, it will

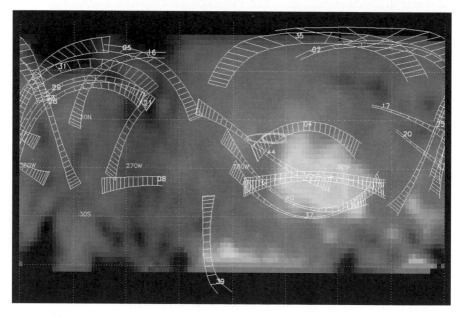

Figure 6.2. *Cassini*'s radar will map all of Titan crudely but will image only about 25% of Titan's surface at its highest resolution. This high-resolution imaging will take place during close approaches in the course of some of *Cassini*'s many passes by Titan over its four-year mission. Deciding which passes is a matter for negotiation with other science teams. This diagram shows the tracks that might be covered under one proposal. The radar tracks are overlain on a version of the HST map and are labelled with the Titan flyby numbers. Their odd shapes are due to a combination of the flyby trajectories and the projection of the map. Some tracks are only half as long as others. On these close approaches, *Cassini* would spend time pointing in another direction to perform a different observation. Image by Ralph Lorenz. (In colour as Plate 11.)

look to the side with five beams that form a line. Dragging this line across the surface will observe a long 'noodle' of Titan – a couple of hundred kilometres across but several thousand kilometres long. Typical sweeps are illustrated in Figure 6.2 and Plate 11. The signals are processed in a clever way, so that pixels only 400 m across can be measured.

With *Cassini,* measuring Titan's size and shape will be taken to new dimensions. The radar altimeter will measure its gross topography and several radio occultations at various latitudes will also contribute data on its shape. Given its rotation and the effects of Saturn's gravity, Titan is expected to be a little flattened along one axis and stretched in another.

Probably the instrument that will capture the public's attention the most is the camera system, or ISS (Imaging Science Subsystem). This comprises two telescopes – a wide-angle camera and a narrow-angle camera. The wide-angle camera is actually a spare copy of *Voyager*'s camera. Both cameras have sensitive CCD detectors and an array of filters. Some of the filters are specially matched to the 0.94-micron 'window' in Titan's atmosphere, to probe Titan's surface. If the haze and motion smear do not interfere too much, features of perhaps a few tens of metres might be visible.

Arranging the observations of all the many experiments on *Cassini* is not an easy task and was made all the harder by some compromises in *Cassini*'s design introduced in 1992 in response to budget cutbacks. In particular, the scan platforms were deleted. These platforms were structures on short booms that allowed the cameras and other instruments to be pointed at specific targets at the same time that the high-gain antenna was pointed towards Earth. Instead the cameras were bolted to the side of the spacecraft, which means that pointing them requires the whole spacecraft to be turned around. This also means that the antenna can't be pointed at Earth while images are being taken; operations have to be arranged sequentially rather than simultaneously.

Phase B – Getting down to business

Two industrial bids were submitted to ESA for the construction of *Huygens*, one led by the French firm Aerospatiale, the other led by British Aerospace, each heading a consortium of European aerospace companies. The proposals were judged on the technical understanding of the problem, the risk and efficiency of the solutions proposed, the strengths of the proposer, and such factors as the division of labour between member states.

In the end, the Aerospatiale bid was selected. Their consortium of course had to include players from elsewhere in Europe, including Deutsche Aerospace for thermal design and the big test and integration job; CASA of Spain would build the structure, Laben and Alenia Spazio of Italy would be responsible for much of the electronics; Logica of the UK would handle the software; Martin Baker Aircraft (famous worldwide for their ejection seats) would be responsible for the descent control system (parachutes and so on). Contractors had to be selected and approved for quality control, ground fixtures, test equipment, shipping, wiring, fasteners, insulation – everything.

The way ESA works – and it is remarkable that it does work – is that contracts on these large projects must be farmed out in the same proportion that ESA receives funds from the governments of its member

Figure 6.3. Technicians at Daimler–Benz Aerospace (as it was then known, now part of EADS) begin final assembly of the *Huygens* probe. The back of the heat shield is visible, as is the top of the descent module. The back cover, on a separate wheeled dolly, is towards the left. Image: ESA. (In colour as Plate 12.)

states, a system known as *juste retour*. This means that the project managers cannot simply pick the best bids for components, or even the cheapest, but have to juggle tasks so that they get everything they need for a tolerable cost, while at the same time making sure that Denmark gets its 3%, Spain its 8% (or whatever the appropriate proportion is deemed to be at the time) and so on. This is a huge headache that to a large extent projects in the USA do not need to grapple with.

RALPH'S LOG. OCTOBER 1990.

I still remember the excitement at receiving the invitation to interview at ESA's main technical centre ESTEC by telegram (possibly the only telegram I've ever had) just as I was taking my final exams in Aerospace Engineering: 'TAKE PLANE TO AMSTERDAM THEN TRAIN TO LEIDEN STOP ROOM AT HOLIDAY INN IS RESERVED IN YOUR NAME STOP ESA VEHICLE WILL PICK YOU UP AT 0830 HRS STOP REGARDS HEAD OF PERSONNEL'. I half-expected the message to self-destruct in 10 seconds.

The prospect of my first job being at the centre of space activity in Europe was an appealing one. I had visited ESTEC the previous summer and was impressed. ESTEC is a few hundred metres from the North Sea and the Dutch girls playing topless volleyball on the beach hadn't escaped my notice either. Happily my interview went well and I joined the project team in ESA in the rainy and windy October of 1990, just before the payload selections were announced – it was, I recall, another month before the industrial contractor was selected. This period was mostly a learning exercise, reading the proposals and understanding the technical issues involved. The ESA project team was about 20 people, a colourful bunch of French, Italian, German, British, Dutch, Swedish.

ESA, by design, is largely a paperwork organisation. It builds little hardware of its own. But when hardware is coming from all over Europe and beyond, it's important. When everything is assembled, it has to work – bolt holes on different chunks of metal have to line up, electrical connectors match up, software interfaces be compatible and so on. All these details have to be documented in excruciating detail before 'cutting metal'.

It felt a little strange for a 21-year-old, straight out of university, to be telephoning professors and telling them that their revised Experiment Interface Document (their 'homework'!) was late, as well as checking these documents to ensure that they complied with ESA's meticulously documented requirements and that what they proposed was feasible. As a native English speaker, I was often called upon to check and recheck documents. The rocket pioneer Werner von Braun once rued 'We can lick gravity but sometimes the paperwork is overwhelming'. My favourite was the 'DRD', or Document Requirements Document, which somehow promised a recursive infinity of paperwork. However, it wasn't all paperwork. I also got the chance to work on exotica like lightning and splashdown dynamics, and the overview I got of the project was wonderful preparation for my Ph.D.

In January 1991, the industrial phase of the project began with a two-day 'kick-off' meeting at Aerospatiale's headquarters in Cannes in the south of France. After a miserable autumn and early winter in the windy Netherlands, the blue skies of the Mediterranean were a welcome change. Over the next few months most of the details of the mission would be hammered out.

Mission plan

The plan in the Phase A study was to launch *Cassini* in April 1996. After flybys of Earth, an asteroid and Jupiter, *Cassini* would arrive at Saturn in 2002. The details of this plan (and crucially, the dates) would change several times in the years to come but, because *Cassini* is so massive that even NASA's largest launcher, the Titan IV, could not

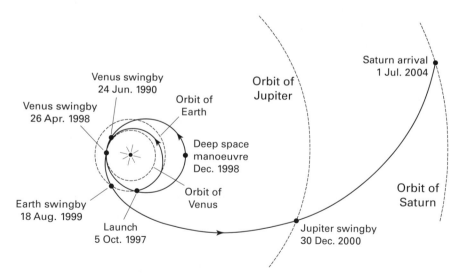

Figure 6.4. The interplanetary flight path of the *Cassini* spacecraft, beginnning with launch on the 15th of October. There were gravity assist flybys at Venus on the 26th of April 1998 and the 24th of June 1999, at Earth on the 18th of August 1999 and at Jupiter on the 30th of December 2000.

project *Cassini* directly to Saturn, most of the incarnations of the mission incorporated an Earth encounter to slingshot *Cassini* out of the inner solar system and a gentle nudge by Jupiter a couple of years before Saturn arrival. In the end, Saturn arrival would occur on the 1st of July 2004.

Cassini's engine will brake it into a long comet-like orbit around Saturn and the craft will climb, ever more slowly, out to several million km. At this point, a couple of months after arrival, its engines fire once again to raise its orbit's point of closest approach to Saturn, the 'periapsis', so that *Cassini* will never again to get closer than about three Saturn radii. This periapsis raise manoeuvre (PRM) also sets *Cassini* up to fly towards Titan.

The first order of business for probe delivery is to get the batteries working. The lithium cells will have built up a thin layer of oxide on their electrodes, which impedes the flow of current (and for that matter keeps the batteries fresh for all those years). By drawing a large current for a short time this layer is burnt off, enabling strong and consistent battery performance afterwards. After the final upload of commands and sequences from the ground, explosive bolts will fire and release the probe which is pushed away by springs. A clever mechanism, called the spin eject device, gives a gentle stabilising spin to the probe as it is pushed away. As the springs push, special low-force electrical connectors come unstuck, literally cutting *Huygens*' umbilical cord to its mother ship.

From this point on, *Huygens* is on its own, like a clockwork toy. It will also get colder than it has ever been or ever will be during its mission. All its systems are shut down, except for three quartz clocks, set to wake it up 18 minutes before it hits Titan's atmosphere. The probe would in fact get too cold for crucial components like the batteries to survive, so it is kept warm with a couple of dozen pellets of plutonium supplying just enough heat to enable *Huygens* to be safely resuscitated. It will coast, asleep and gently spinning, for 22 days, during which time Titan will sweep around Saturn one and a half times to appear underneath *Huygens* at their appointed rendezvous, initially planned for the 27th of November 2004 but now (see later in this chapter) set for the 14th of January 2005, three Titan orbits later.

Cassini will in the meantime reorient itself and make another burn of its engines. This burn slows *Cassini* down with respect to *Huygens*, so that *Cassini* arrives several hours later (the so-called Orbiter Delay

Figure 6.5. An impression by ESA artist David Ducros of the *Huygens* probe shortly after release from the *Cassini* spacecraft. When the event actually takes place, Titan will be farther away than depicted here. The artist has also been unable to resist the temptation to show the rings ajar; in reality, they would appear edge-on. Artwork: ESA. (In colour as Plate 13.)

Figure 6.6. (a) (opposite) The original plan for *Cassini*'s trajectory between its initial arrival at the Saturn system in June 2004 and the arrival of the *Huygens* probe at Titan in November 2004. In the new plan, the first orbit is smaller, encountering Titan in October 2004. *Cassini* makes 2 ½ orbits before the probe is released.

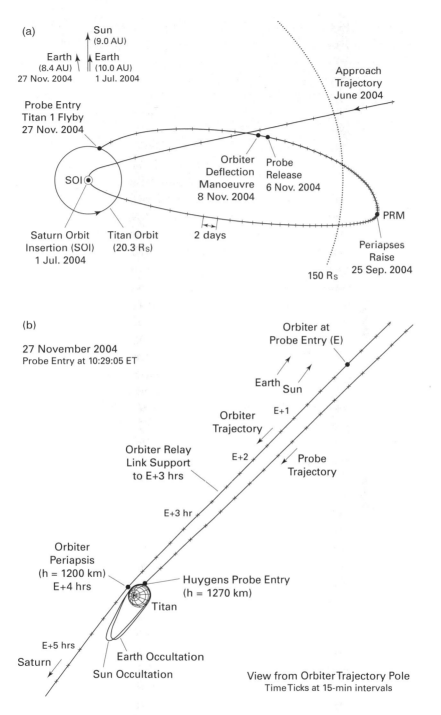

(b) Details of the original trajectories of *Cassini* and *Huygens* just before and during the probe's entry into Titan's atmosphere and descent to the surface. In the revised plan the orbiter is much further away (periapsis > 50 000 km).

Time, ODT) and can act as a radio relay. It also displaces *Cassini* to one side, so that the orbiter flies by, rather than into, Titan.

Cover me – I'm going in

Huygens wakes up a few thousand kilometres above Titan, travelling at a speed of about 6 km/s. The computers boot up and various instruments are turned on. Most electronic systems perform best at constant temperature. A particular example is the ultrastable oscillator on the Doppler wind experiment, which takes a short while to settle down to its most stable frequency.

The first science *Huygens* does begins at about 900 km altitude, when the thin atmosphere begins to tug noticeably at the craft. Noticeably means the detection threshold of the accelerometers of the HASI experiment: a few millionths of Earth's gravity (i.e. 10^{-6}g), when the atmosphere is about a billion times thinner than at Titan's surface. *Huygens* is slowed by the drag on a blunt conical heat shield, 2.7 m across. The shield is protected by a 1–2-cm-thick layer of ceramic tiles (a non-reusable version of the tiles used on the Space Shuttle, derived from French missile technology.) As *Huygens* plunges deeper into the atmosphere, the heat and drag from the air tortured by the probe's passage tears away the shiny foil coating from the outside, the multilayer insulation that protected *Huygens* from the fierce heat of the Sun while *Cassini* was at Venus. Now at peak heating, at an altitude of about 400 km above the surface, *Huygens* has barely slowed and the air in the shock wave in front of the probe is heated to 1400 K. In fact it glows violet because the carbon and nitrogen atoms from the methane and nitrogen molecules in the air together emit violet light. Were it not blazing with several hundred times the heat of the Sun, it would doubtless be pretty to look at. The intense glow from the heated air requires the back of the probe to be covered with insulation.

A few seconds after peak heating, *Huygens* feels its maximum g-load at around 250 km. Although *Huygens* is moving more slowly than at peak heating, it is in denser atmosphere now and feels a deceleration of about 15g. This is in fact quite modest by planetary probe standards: the *Galileo* Jupiter probe endured about 250g and the *Pioneer Venus* probes had to withstand 400g. Because the arrival velocity of *Huygens* is quite low and the atmosphere is so extended, Titan is a gentle cushion. About a minute after peak deceleration the probe has slowed to about 350 m/s, or about one and a half times the local speed of sound, at an altitude of about 170 km. The 'sound barrier' applies to bodies breaking the speed of sound in either direction and the fairly flat *Huygens* probe would begin tumbling as it passed Mach 1. So at Mach 1.5, at 170 km altitude, the action begins. A mortar on the back of the probe shoots a bundle through a weak patch in the back of

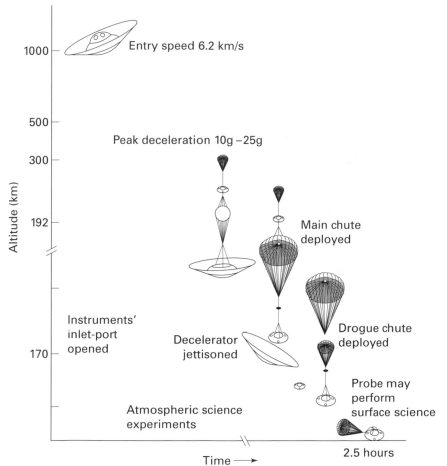

Figure 6.7. The sequence of events between *Huygens*' initial encounter with Titan's atmosphere and its impact on the moon's surface. As the probe slows down, a small parachute will be released, which will then deploy the main parachute. Once the main parachute is fully open, the decelerator shield will be jettisoned and the probe will drift downwards. About 40 km above the surface the main parachute will be jettisoned and a smaller drogue chute will carry the probe the remaining distance. Scientific data will be transmitted continuously to the *Cassini* orbiter during the 2.5-h descent and relayed to Earth later. If the probe survives its impact at about 10 miles per hour, it may continue to transmit data to *Cassini* for up to a further 30 min.

the probe's aft cover. This bundle blossoms into a small 'drogue' parachute, which holds *Huygens* stable and pulls the back cover off the probe. As the back cover comes off, it pulls out the main parachute, a huge circular canopy some 8 m across. After the chute inflates, the heat shield falls away.

This altitude merits a moment's reflection. At 170 km altitude on Earth, objects are in space. The Space Shuttle usually flies at about 350 km. Aeroplanes usually fly lower than 15 km altitude. Titan's atmosphere is hugely distended.

As the main chute opens, the probe decelerates quickly to a steady descent rate of about 50 m/s. A cover that has protected DISR from material coming off the heat shield is released and flies off on springs. Similarly, caps on the inlet pipes of the GCMS are broken off by explosive actuators and allow gas sampling to begin. Two small arms bearing HASI's electrical field sensors swing out and lock into position. Small vanes mounted around the edge of the probe keep up a slow spin under the parachute to pan DISR's camera around in all directions; the line to the parachute has a swivel to prevent it twisting up.

The size of the main parachute was dictated by the need to pull the probe away from the heat shield safely. But if the probe continued to descend under the main parachute, it would take some 5–8 h to reach the ground, by which time, the *Cassini* orbiter would have disappeared out of sight before the surface data could be taken. So after 10 minutes the final explosive bolts are fired to release the main parachute. A smaller stabiliser parachute allows the probe to descend rather more quickly.

The ACP experiment sucks atmosphere in through a filter, trapping aerosol particles. A small oven breaks them down so their composition can be analysed by the GCMS. At 40 km and below, methane condensation is possible and, at the cold altitudes above about 14 km, a thin film of methane frost may form on the probe, although not enough to disturb its aerodynamics. (Icing is a problem on aeroplanes because the ice build-up can disrupt the delicate lifting airflow over the wing, but *Huygens* is a 'bluff' body that doesn't generate lift.) Below 14 km any condensed methane will be liquid and the probe may fall through clouds or even rain.

As the probe, descending at about 5 m/s, nears the surface, as determined by a radar altimeter onboard, the instruments adapt their operations to maximise the scientific return. At an estimated two minutes to impact, the SSP experiment primes itself. Its acoustic sounder – like a small sonar – begins sending out pings at a high rate. A lamp will illuminate the surface in the last few tens of metres of descent allowing DISR to take a surface spectrum unaffected by the atmosphere.

Anything could happen at impact. We have no idea what the surface is like mechanically. Maybe it's covered in thick dust-like organic material and the probe will sink in several metres, as was once thought might be the fate of landers on the Moon. Or maybe it will be a hard glazed ice surface. If it is, the probe's lower surface will crumple and it will die. More likely perhaps is an intermediate surface, covered in a sludge or snow-like deposit, and *Huygens* will keep operating. Another possibility is that the probe will splash down into a lake of liquid hydrocarbons. Because nothing is known about the Titan surface, it was not feasible to design the probe so it would definitely survive – certainly not for a reasonable cost. But for most surfaces, there is a good chance that *Huygens* will keep transmitting data from the surface.

One concern was whether the probe would float and, if so, whether it would it do so with the parts of the probe requiring access to the liquid (like the Surface Science Package) below the 'waterline' but with the camera above it? It will. In fact the period with which it bobs up and down can be used to infer the density and the composition of the liquid.

If nothing breaks at impact, the probe will continue to take data but it should be borne in mind that, unlike *Mars Pathfinder*, the probe is not designed for operations after it has landed. There will be pictures but they will be limited to wherever the camera points when it comes to rest, whether at a rock, or blank sky, or maybe even the underside of the parachute. There's a chance that some surface material may be vaporised by the heated inlet of the GCMS so that it can be analysed. In a liquid the acoustic sounder on the probe may pick up an echo from the bottom and hence determine its depth.

Ultimately, one of several things will happen. The batteries will run out or, as the probe cools down, something – perhaps the batteries, perhaps the radio transmitter – will stop working. *Cassini* flying overhead will disappear out of sight, although somewhat before then the accuracy with which the antenna on *Huygens* is pointed at *Cassini* will become so poor that the data link will die out.

After listening out for the probe, the *Cassini* orbiter will slew around to take pictures and spectra of the areas as near the landing

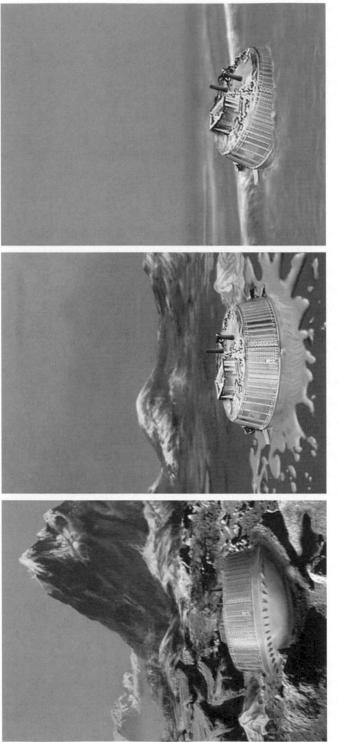

Figure 6.8. An artist's impression of three possible scenarios for the landing of the *Huygens* probe on Titan. From left to right: the 'crunch' (on solid terrain), the 'squelch' (in slush), and the 'splash' (in liquid). Artwork by James Garry. (In colour as Plate 14.)

site and as close to the *Huygens* descent time as possible. These data will be key in understanding future Titan observations from *Cassini*. Several hours after the Titan flyby and the end of the probe mission, *Cassini* will slew over to point at Earth and will disgorge its hard-won load of data from the *Huygens* mission.

1% inspiration, 99% perspiration

All that at least is the theory and the goal everyone had in mind during the seven long years of hard work leading up to the launch. There is a fundamental tension in a scientific space project. In a sense the scientists are the 'customers'; they naturally want every facility and convenience and wish to be allowed to supply their experiment as late as possible, as heavy and power-hungry as they can, built and operated according to standards and techniques with which they are most familiar. The scientists are not responsible to ESA. The money to build their experiments comes from their own national governments. This is a significant difference between Europe and the USA, where both the spacecraft and the scientific funding are provided by NASA. Industry wants as much money as it can get but is contractually obliged to ESA. To make their life easy, the experiments should be supplied early and the specifications and requirements fixed ('frozen') early, so they can get ahead and cut metal. Sandwiched in between is ESA, with the difficult job of cajoling everyone into using the same standard and organising delivery dates that are achievable for the scientists but acceptable to industry, and vice versa. Some of these details, like the difficult debate over whether to use rectangular or round electrical connectors, took many exchanges of documents and hours of painful meetings to resolve. Sometimes unhappy compromises had to be forced on scientists or the contractors, the tough job of the project manager. *Huygens*' veteran project manager, British-educated Hamid Hassan, seemed sometimes to relish his role of being the 'bad guy' and, if you were summoned to meet him, often to cringe across the table through a cloud of evil-smelling cigarette smoke, it was not usually to receive happy news. Still, he got the job done, which was why he was in charge.

Cassini carries another source of tension: the cultural and procedural differences between the two sides of the Atlantic. The US-supplied experiments on the probe naturally had to comply with ESA standards, such as using metric measures on engineering

Figure 6.9. The special model (SM2) of the *Huygens* probe lying in the snow in northern Sweden after its parachute drop from a large helium balloon at an altitude of 37.5 km. The model was barely damaged and equipment on it to measure the parachute function and dynamics unexpectedly continued to operate after impact. This outcome raised hopes that the real *Huygens* may continue to operate after an impact on Titan. Small square panels protruding from the side of the model are radar altimeter antennae. ESA image.

drawings. When handling lots of paperwork, even the difference between US letter-sized paper and the European A4 size can become a pain! Despite the deep-seated tensions, the years of working together forged many solid friendships. Particularly with a long-term project like *Cassini*, relationships develop, almost to the point where being a team member is like belonging to an extended family.

Over the years numerous issues had to be dealt with – analyses showing a cover would be needed for the DISR windows, poor linearity of the radar altimeter, the possibility of icing, difficulties obtaining the valves for the GCMS, worries about how long the ultrastable oscillator would take to settle down, poor performance of foam insulation, and so on. Dozens of issues like these had to be dealt with – either fixed outright, sometimes by people working 80-hour weeks, or their impacts fully quantified and accepted.

Sometimes entirely new tests had to be devised and performed. One special test was carried out in northern Sweden, where a high-altitude balloon facility was available to drop a model of the probe from an altitude of 37 km and test the complicated parachute sequence. This kind of test is always traumatic. Not only is it yet another item for the project to pay for and supervise but there is also the consideration of what happens if something goes wrong. Is there time to fix it, if indeed something specific to the probe (rather than the test) goes wrong? And if it was a test problem, is there enough time or money to do it right? Happily the drop test went well and, as a bonus, the systems kept working after the probe landed in the snow, raising the hope that the real probe may stand a good chance of working after impact.

As on almost all space projects, everything seemed to get heavier and measures had to be taken, like reducing the number of batteries. And there is always intense pressure on budgets. While it is true that 'engineering is the art of doing with one dollar what any damn fool can do with two', sometimes the budgetary pressures (especially on the scientists whose funding usually comes from excruciatingly parsimonious national agencies) made life difficult. In the first few years of the project, the British experiment teetered on the brink of unfeasibility but its canny PI John Zarnecki somehow kept it all together.

RALPH'S LOG. 1991–1994.

The Huygens probe weighs some 320 kg but a particular 14 grams of it will always be dear to me. It is rare indeed that individuals get the chance to build something with their own hands that leaves Earth. The little bit of Huygens to which most of my Ph.D. was devoted will be the first part of the probe to touch the surface of Titan.

After my year in ESA, I saw that the scientists seemed to have more interesting challenges than the engineers, so I returned to the UK to begin a Ph.D. at the University of Kent with John Zarnecki as my supervisor. Although the transition from a well-paid, jet-setting engineer to student wasn't easy, the chance to actually build part of the surface science package was too good to miss. My undergraduate training as an aerospace engineer, and a simple fascination with the idea of splashing down into an ocean of liquefied natural gas, drew me to the measurements to be made of Huygens' impact on Titan.

Impact measurements have long been made by spacecraft. Indeed some of the earliest proposed lunar missions (back in the days when it was feared there were seas of dust into which unsuspecting astronauts might sink) were only spheres with accelerometers, so that the hardness of the surface could be measured. In the assessment study and the early Phase A study that followed it, the European Space Agency recognised that the Titan atmosphere probe (as it was then called) might survive for a few minutes or so on the surface and that it might be worthwhile to include at least a small element of the payload to make measurements of the surface properties. An accelerometer was identified as one possible (and easy) sensor. The deceleration on hitting the ground could tell at least whether the surface was solid or liquid.

Accelerometers sense the rate of change of speed. Many modern cars have them to activate airbags. They are often used on planetary probes (including Huygens) for atmospheric measurements. The acceleration relates directly to the drag force as the spacecraft plunges into the atmosphere, which in turn depends on the atmosphere's density. Knowing the arrival speed, the density profile of the atmosphere can be calculated from the deceleration record. This technique was pioneered by Alvin Seiff of NASA's Ames Research Center. He applied it to the Vikings, Pioneer Venus, Galileo and Mars Pathfinder. In a similar way, the deceleration at impact would be diagnostic of the surface material: but how?

The theory for splashdown dynamics was quite well documented in NASA reports from the 1960s, where the Mercury, Gemini and Apollo manned capsules returned to Earth by splashing into the ocean. I found by applying the relevant calculations to Huygens in a hydrocarbon ocean that the deceleration loads should be quite modest – about 15 or 20g – not much more, in fact, than during the hypersonic atmospheric entry. The entry accelerometers, part of the HASI experiment, should do a fair job of recording the splashdown.

However, landing on a solid surface would not be as simple to interpret. Spacecraft are not like cannonballs. They have to have a fair amount of empty space in them. Huygens is a thin metal shell, with equipment boxes bolted to a honeycomb platform and enough empty space to allow the hundreds of cables and connectors between the boxes to be assembled. If Titan's surface is so hard that the probe deforms, an accelerometer would learn less about the surface than about the probe itself!

I decided that what was wanted was not an accelerometer at all. We were interested in the force applied by the surface material. Therefore, we should measure the force directly on something rigid – a mast sticking out of the bottom of the probe. The sensor/mast assembly was called a 'penetrometer'. I spent my first few weeks in Canterbury looking at catalogues of force sensors. They could indeed by procured but were expensive. We therefore looked into building our own. Some very sound advice was given to me by Tim Stevenson, then the SSP programme manager. Rather than optimising the design on paper, he suggested I should just build something quickly and get familiar with how the hardware worked as soon as possible. We would make the force sensor out of a piezoelectric ceramic, a material that generates a charge when it is compressed. (It is also used in gas lighters to generate a spark when the handle is squeezed.) A bonus was that Zarnecki and crew at Canterbury were familiar with piezo-electric sensors, having used them as microphones to detect the barrage of dust impacts on the Giotto spacecraft that flew through the coma of Comet Halley.

Within a couple of months we had put together a mock-up. To simu-late the Huygens probe, which will have a mass of about 200 kg after the heat shield has been discarded, we used a set of steel weights – around 5 kg. The mass difference wouldn't matter as long as the mock-up's speed didn't change much while the mast, which was around 20 cm long, penetrated the surface. The impact speed, at least, was easy. Knowing the size of the parachute and the approximate density of Titan's atmosphere, this would be around 5 m/s, about the speed that a calculator dropped from a desk hits the floor. This is con-venient for testing and calibrating the sensor. There would be no need for giant catapults or guns to shoot the device into test surfaces; we could simply drop it.

The first crude versions of the sensor broke after a few drops – good enough to get some initial test data but too unreliable for the real thing. The signal quality wasn't very good. The fact that my test sand was damp didn't help and, to save money, I was using a slow and awkward early 1980s microcomputer to log the data. I presented some of this very early data at a meeting in Edinburgh that April. I could at least discriminate between drops into a bucket of sand and drops into a bucket of ball bearings. However, we were on the way to a better design.

We tried a conical tip on the penetrometer but found that the signals from this design were weak and contaminated with a lot of ringing; the impact generated a lot of side forces on the cone. A spherical tip did rather better than the cone, although it still rang somewhat. Seiff had joined the HASI consortium and paid us a visit. It was encouraging to have a veteran take an interest in my little bucket of sand, make sug-gestions and explain why the round tip worked better.

By now, summer 1992, with a budget at least assured if meagre, the rest of the SSP project was gearing up for action and much of this effort was on the electronics side. I spent three months at the International Space University (ISU) summer school in Japan and, when I returned, there was a new set of computers in the lab, required for communicating with the experiments and archiving test results. These made taking data from the penetrometer vastly easier. Sometimes going away and leaving things alone is the best policy.

SSP had grown and was beginning to look like it might become too heavy. Mark Leese, the new SSP programme manager, asked whether we could make the penetrometer smaller – it was some 28 mm across. I couldn't think of a reason why not, so we halved its size. This, and the use of titanium for the tip, saved tens of precious grams, although the sensor became very difficult to assemble, requiring much threading of wires and gentle coaxing to put together.

Another addition was the inspiration of Zdislav Kryzinski, an electronics engineer seconded to the SSP project from Poland (one of Zarnecki's resourceful ways of dealing with SSP's manpower shortage). There was an ESA requirement to verify sensor function during the in-flight checkouts in the course of the 7-year cruise to Titan. Kryzinski's idea was that we could put an extra electrode on the piezoelectric sensor. Sending a voltage pulse to this electrode would make the sensor deform slightly, which would be picked up as a voltage on the regular sensing electrodes. Although it prompted many jokes about 'stimulating my penetrator', it worked (more as a capacitor than electromechanically, but a source of reassurance nonetheless).

There were occasions when nothing would work, throwing me into days of despair. Alternative and more lucrative careers like stockbroking began to look appealing. But then I'd remember the bigger picture, how this crappy little sensor fit into the great adventure of planetary exploration and how lucky I was to be doing this, and eventually the problems would get fixed.

Now that we were converging on the final design, we could invest more effort in seeing how the sensor would work and we also had to think about how to calibrate the sensor. Piezoelectric sensors are very good at measuring short signals but terrible for measuring steady forces. So calibration was not as simple as putting a weight on the sensor and measuring the voltage. Somehow we had to generate a force pulse.

What we ended up with was a Meccano assembly with a swinging arm that could be raised to one of several end stops and dropped. At the end of the arm we could place a lump of material, such as a piece of plastic or rubber. This fell at a known speed onto the sensor, which was held fixed.

It sounded and looked primitive but the shape of the force pulse produced was very smooth and very reproducible. From one drop to the next the signals produced were virtually identical. One big question was, 'How well would the sensor perform at Titan's super-cold temperatures?' In principle, it should work well. All the materials had been chosen for their temperature stability. The acid test was to chill the sensor with liquid nitrogen and perform the calibration again. It worked. The signals produced were about a factor of 2.5 smaller than at room temperature and it took a few weeks of thinking to work out why. It turned out to be a combination of the change in electrical properties of the piezoelectric material (which were in fact rather more than the manufacturer's specification) and in the mechanical properties of the plastic used to insulate the sensor.

There were some other details to be worked out, none of which was too problematical. We had to be sure that radiation wouldn't damage the material, so some piezoelectric disks were sent to a reactor to be irradiated. Also, we had to design a foil or mesh cover so that electrical charge from atmospheric aerosols or rain on Titan wouldn't cause electrical discharges.

On one occasion, a colleague, Phil Daniell, berated me for having a large magnet in my bucket of sand. I had been trying to measure the effect of ball bearings in the sand and the easiest way to sift them out was with a magnet. He felt this was a hazard and that vital computer data might be lost if a floppy disk was corrupted by the magnet. I pointed out that anyone who put a computer disk in a bucket of sand deserved everything they got but acquiesced to the extent of mounting a sign, saying 'Danger – Sand and Magnets' on the bucket.

A couple of students, Mick and Greg, helped with some of the tests as undergraduate projects. We dropped the sensor with a set of weights down a long drainpipe to control its impact angle and speed. We tried different materials as well as the sand. I headed to a craft shop in Canterbury to get modelling clay and grabbed some buckets of gravel from the construction site of a new pathway on campus. Clay gave a nice steady signal, as you might expect for a very viscous fluid. Sand had a signal that ramped up; presumably the sand gets more tightly packed at greater depths. Gravel had a very spiky signature, with the spacing and height of the spikes depending on the size of the gravel.

I was very pleased. My little sensor looked like it would measure something interesting and useful. We had the final components made up. I had by now written up my Ph.D. thesis and was getting ready to leave Canterbury and to move to the USA. I wanted to build the sensor that would actually fly. We got everything ready for one afternoon. I would do the mechanical assembly, while Trevor Rees, a technician in the physics department's electronics lab, would do the electrical assembly. (The quality control regulations require that soldering be performed by someone who has followed a special soldering course.) The assembly was fiddly, even more so with gloves. The worst part was tightening the titanium bolt through the sensor. If the components slid around too much during the tightening, the tiny wires inside would

Figure 6.10. The 'penetrometer' – *Huygens*' fingertip. The small titanium hemisphere will by the first instrumented part of the *Huygens* probe to make contact with Titan's surface if all goes well. Assuming the probe does not land in liquid, the piezoelectric sensing element (the light-coloured disk between two darker washers under the hemispherical head) will generate a signal proportional to the force applied to the tip at impact, from which the mechanical properties of Titan's surface can be deduced. Image by Ralph Lorenz.

be sheared and perhaps broken. The bolt had to be tightened to a specified torque.

We made a batch of five. The first one wasn't too good but was useful practice. The best one was to fly. Two others were good enough to be considered flight spares. One of the others cracked during tightening. We bagged the good ones up, after some brief calibration measurements. My work was done.

Models and mating: a match made in heaven

All across Europe and the USA, there were dozens of stories like the one above – some of novices like myself, others of veterans, in industry, academia, NASA and ESA. Many parts of *Cassini* and *Huygens* had never been used on spacecraft before. The management and engineering procedures that allow the activities of people dispersed worldwide to come together is perhaps the greatest legacy of the *Apollo* era, not the often-mentioned (but incorrect) spin-off of non-stick frying pans. It is an amazing feat.

When the *Cassini* orbiter and the *Huygens* probe finally met in April 1997, it was not exactly a blind date. Their perfectly formed but brain dead cousins (the respective parts of the 'structural thermal and pyrotechnic model', mercifully abbreviated to STPM) had already met to establish that bolt-holes would line up and so on and that all that paperwork had been useful! The assembly was then tested on a vibration platform, to check that no undesirable resonances, which should in any case have been identified already in purely virtual computer models, had been introduced. Even closer relatives – ugly but intelligent siblings known as 'engineering models' – had linked spaghetti-like cables to verify all the electrical connections and signal protocols, whether the software would work correctly, and that the different experiments did not interfere with each other.

The probe is largely self-contained and unresponsive before it is deployed into Titan's atmosphere but the sensitive orbiter feels its presence. Although careful selection of nonmagnetic materials and the use of compensation magnets where magnetic materials are unavoidable (as in the pump in the aerosol sampler), minimise the magnetic field of the probe, the *Cassini* magnetometer is easily able to sense the probe. Similarly, the radio and plasma wave system can detect all kinds of weak radio noise associated with the probe's computers and voltage converters.

Bringing the hardware together requires paperwork for more than just the hardware itself. Handling procedures, the layout of test facilities, staffing plans, security information for the personnel involved: all this must be worked out in advance. The design of test fixtures (like the big jig used to wheel *Huygens* around and mate it to the orbiter) and shipping containers must all be done in advance. Shipping paperwork too can be a headache. For example, a critical component of the surface science package, a small tilt sensor sent from the USA, was held up in London's Gatwick airport for days while the customs paperwork got sorted out: a frustrating and unnecessary delay for the project manager who was fretting about an imminent deadline. A non-technical but illustrative episode happened in 1997. ESA had printed a load of special *Huygens* T-shirts to give to the science and engineering teams at the launch. But, the T-shirts had been made in Turkey, which for some reason meant the US customs service would not allow their entry. No T-shirts for us.

A particularly frustrating experience has been the over-vigorous application of ITAR (International Traffic in Arms Regulations) in the USA, particularly following the case of a Chinese scientist at a US nuclear facility accused of stealing secrets. Many aerospace components like accelerometers and, perversely, even the software used to predict *Cassini*'s trajectory through space and its orientation, are classified as 'munitions' and their access by non-US nationals is restricted. This is a tremendous inconvenience to European scientists. These kinds of problems work both ways, however. An error in the *Huygens* radio system design (see later) might have been picked up before launch and corrected if critical details were not stuck in an Italian factory to which JPL experts did not have easy access.

The politics of Battlestar Galactica

Cassini is a monster. In the early 1990s, recently appointed NASA chief, Daniel Goldin, stated publicly that *Cassini* would never have been started under his administration. At that time, the NASA planetary program was becoming stifled by the few, large missions on the books. There were *CRAF Cassini*, *Magellan*, *Galileo*, *Mars Observer*, and not much else. Part of the problem was that, to justify the existence of the Space Shuttle, planetary missions planned in the late 1980s had to be large to fill the Shuttle's cavernous cargo bay. Large, expensive missions take a long time and a lot of money to develop. For the outer solar system, that is unavoidable but for Mars, which can be reached in only a year or two, it is a restrictive policy. Goldin's new catch phrase was 'faster, better, cheaper'.

The logic behind this philosophy is sound from the technological point of view. *Cassini* indeed represented a gamble: two billion eggs in one basket put atop a rocket. All rockets have about a one in thirty chance of failing. However, at the end of the 1990s, NASA had been burned by a salvo of Mars failures and the faster–better–cheaper approach tarnished somewhat. It perhaps got taken too far but it was definitely the right way to go given where things stood in relation to missions within the inner solar system.

However, the outer solar system is a different ball game. A small, focused mission runs the risk that, before it reaches its target, its objective may be superseded by new data from the ground or better ways are developed of performing the one measurement it was intended to make. It has been calculated that it would take nearly a billion dollars to put even a ball bearing in orbit around Saturn. To double the cost by increasing the science return a millionfold looks like a sound strategy. Wide political support and broad scientific scope are needed because the time to get there is long, particularly compared with the four- or five-year election time scale of politicians. A bold politician who advocates an expensive new project is unlikely to get the credit when it arrives ten years later.

Cassini's observations are made many times more useful by the synergy of simultaneous observations: seeing the same thing in different ways. Its wide range of scientific objectives, to explore not only Titan and the other satellites but also Saturn, its rings and magnetosphere, assured a broad base of scientific support. If *Cassini* had been broken up into a number of smaller missions, each

Figure 6.11. Flight mechanics from JPL lower the *Cassini* spacecraft onto its launch vehicle adapter. The adapter was later mated to the Titan IV/Centaur launch vehicle. NASA image. (In colour as Plate 15.)

one might have been more prone to cancellation, or a death of a thousand cuts.

As it was, *Cassini* only survived by the skin of its teeth. In 1992 Congress almost forced NASA to decide between building the

International Space Station and flying *CRAF/Cassini*. Normally, planetary exploration would lose this battle and *CRAF/Cassini* would have been cancelled. However, the international dimension of *Cassini* proved pivotal. The USA's reputation as an international partner would have been tarnished forever if it had unilaterally cancelled the project. After a concerted lobbying effort, *Cassini* was saved.

There was a price to pay. *CRAF* was cancelled, meaning that NASA would not get two missions for the price of one and a half (since *CRAF* and *Cassini* had many elements that could be developed in common to save money overall). *Cassini* also had to slim down. Crucially, the scan platforms that would have allowed its instruments to point in different directions were cut and many instruments were painfully 'descoped'. Software developments that would give *Cassini* its full capabilities at launch were deferred.

Cassini was almost cancelled again in 1994/5 but again the international dimension saved it. When it finally reached the launch pad it was the most massive interplanetary spacecraft launched by the USA and, apart from the ill-fated Russian *Phobos* spacecraft, the largest interplanetary spacecraft launched by anyone. Carrying a dozen instruments, *Cassini* is a superlative spacecraft.

RALPH'S LOG. OCTOBER 1997.

If all went well, Cassini–Huygens was to be launched from Cape Canaveral sometime between 6 October and December 1997. If some development problem arose and the launch slipped beyond December, it would not be able to 'catch' a Venus flyby and would take an extra 2 years to reach Saturn. This would not be good; the longer a mission spends in space, the greater the chance some mishap might befall it. Further, arriving two years later would mean that the rings would be more edge-on to the Sun, making many ring observations difficult. Two more years of plutonium decay would diminish the available power on the spacecraft, making it harder to perform simultaneous observations with many instruments. Finally, two more years in space would mean two more years of operating cost, without increasing the science return significantly.

So the pressure was on. The project team was put under siege by environmental protesters, the more extreme of whom saw Cassini with its plutonium power supply as part of a government conspiracy to introduce nuclear weapons into space and had threatened to break into the launch complex to try to prevent the launch. They filed law-

suits claiming that NASA threatened their lives due to the infinitesimal risk of a launch accident releasing plutonium. Under US law, they needed only to find a sympathetic judge somewhere. In fact the most significant threat was a lawsuit filed in San Francisco.

This is a very touchy matter, as are many issues relating to events with low probabilities but huge effects. One statistic the protesters tried to scare the public with was that if the plutonium were distributed worldwide, it could deliver a toxic dose to billions. However, applying the same logic, anyone suffering from the flu who sneezed could be accused of attempted genocide.

The launch was a pivotal point in the project. For some it would mean a crescendo of activity as they performed last-minute tests at the launch site, since this would be the last chance to fix anything. For everyone, it would be something of a celebration – the culmination of years' of work and the first chance to see with their own eyes some of the hardware. Most of all it would be a relief. After all the usual technical difficulties of a spacecraft development, years of lobbying and campaigning to make sure the mission wasn't cancelled, we would know that Cassini was safely on its way. Many brought their families, filling the hotels in Florida's Cocoa Beach

One last-minute technical problem did arise in the weeks before launch. The Huygens probe is warmly wrapped up for its brief descent in the frigid Titan atmosphere, so cool air was blown into the interior for the long tests on warm Earth. Somehow, the airflow on the cold air line was set too high. The blast of air flapped some of the insulation blankets to pieces, blowing little particles of foam around the interior. The damage would have to be inspected but this would mean dismounting the probe and opening up its entry shield. The launch would have to be delayed!

This was of course the right thing to do. Seven years hence, if something didn't work properly, the recriminations would have been terrible. For the project engineers at the Cape, it was one more crisis to deal with as calmly and professionally as possible. For the environmental protesters it was an opportunity; 'look at the incompetence of these people', they howled. For most of us, now distanced from the hardware and powerless to help, it just meant a scramble to change our flight reservations.

The problem was fixed and the launch rescheduled for the early morning of 13 October. To lighten the tension, the science teams set up a soccer match between Huygens and Cassini, rather an unfair choice of game, given the preponderance of Europeans on Huygens and the much greater popularity of that game in Europe.

The crowds gathered on the evening of the 12th, boarding the buses to be taken to the viewing site, looking out across the sea to the launch site. There was a festive, but tense, atmosphere. Lots of old acquaintances and friends bumped into each other. Lights from the open-air TV studio blazed. A TV set up at the waters' edge showed a close-up of the launcher and a large clock showed the countdown. But mere minutes before the launch was supposed to occur, the message came across the radio net that a computer glitch meant the launch was scrubbed. Disconsolate, we trooped back to our hotels in silence, reaching our beds at around 4 a.m. I got up again at 8 a.m. for a meeting of the Surfaces Working Group, to discuss mapping tools and opportunities for viewing satellite surfaces in the new version of Cassini's orbital tour. To my chagrin, the meeting had been delayed by an hour and in the dark disappointment at the viewing site, word had not been spread.

The launch was rescheduled (again at an ungodly hour) on Wednesday the 15th. The viewing site was less crowded, in part because some people had headed home, unable to prolong their stay. Others had heard of better, or at least more accessible, viewing sites. (Many of the hotels had a decent view.) The mood was rather different – somewhat more sober, or cynical perhaps. As we waited, chatting, a few alligators swam idly in the waters in front of us. The countdown progressed nominally and tension started to mount. In otherwise clear skies, a big cloud drifted over the pad.

I sat with my wife and several of the SSP team: Mark Leese, Phil Daniell, John Zarnecki and John Geake, who had designed an elegant part of SSP to measure the refractive index of Titan's seas, if we landed in liquid. Tragically John Geake died about six months after the launch. Somehow it came up in conversation that, between my leaving Canterbury and the launch, my precious flight model sensor had been broken! Apparently in the process of being attached to the rest of SSP, the wrong part had been tightened and it had cracked. The spare was now on Cassini! I hoped I'd be able to find the calibration records of the spare . . .

As the countdown got to the last 10 seconds, the crowd began to chant the numbers. My wife squeezed my hand. Cameras whined near-inaudibly, charging their flashes which would of course be useless. Flames billowed around the base of the rocket – had it caught fire?

No. A moment of silence was deceptive and short-lived. The rumble of the normal ignition took a couple of seconds to propagate through the ground and the sea. Soon the rocket lifted off safely. A few hapless flashes went off and whoops and cheers volleyed from the crowd. As more of the rocket plume was exposed, it became brighter. Then it van-

Figure 6.12. The launch of the *Cassini–Huygens* mission lit up a small cloud from the inside, which glowed like a Chinese lantern in the dark as the rocket headed skywards. Image: NASA.

ished into the cloud above the pad, illuminating it from inside, like a Chinese lantern. It was a pretty spectacle.

The direct sound of the launch finally reached us with a roar, perhaps less loud than some of us expected. The rocket arced over the sea and presently the two solid rocket boosters fell away, tumbling slowly. We were off!

Touring the Saturn system

One of the main activities of the late 1990s for the *Cassini* teams was to devise and select the tour, the complex set of orbits that *Cassini* will perform. From an infinity of possibilities, somehow the most scientifically rewarding path had to be found. *Galileo*'s tour around Jupiter was much simpler – only two-dimensional, in the plane of the orbits of the Galilean satellites, all of which it utilised to alter its trajectory. For *Cassini*, Titan is the lynchpin of the tour. Only Titan's gravity is strong enough to change *Cassini*'s orbit around Saturn appreciably, so the tour is constructed around a series of Titan flybys. But this means *Cassini*'s tour can be three-dimensional, opening up a bewildering but rich array of possibilities.

Back in the project's prehistory, a possible tour called 88-01 had been constructed. This was the tour illustrated in the Phase-A report and announcement of opportunity. It had 36 Titan flybys, two of Iapetus and one each of Enceladus and Dione, together with a suitable balance of ring observations, magnetospheric orbits, Saturn occultations and so on. An intermediate iteration called 92-01 added some improvements but, in 1994, the planners at JPL really got down to business and developed a series of tours to 'explore the parameter space' – basically trying a variety of combinations of strategies to see what the possibilities were – before narrowing down on the most promising choices. One set, exploring various permutations was named A1 to A8; then a twist on these was B1 to B8 and so on – to F9 or so – almost 50 in all.

RALPH'S LOG. DECEMBER 1994.

The scientists were initially not very well equipped to evaluate which of these tours were better or why. Only the broadest aspects could be studied. When I moved to Arizona, my involvement had been only with the probe, so I knew and understood little about the tour. However, I had always loved orbital mechanics and, in particular, the elegance of changing a spacecraft trajectory by making a planetary flyby. Working with Jonathan Lunine, I became associated with the radar experiment on Cassini and was tasked with evaluating the tour possibilities. I counted Titan flybys and their distribution by making dots on squared paper – one row for each tour. A dot in a column corresponded to a flyby at a given range of latitudes. A good spread of dots meant broad coverage – good news for radar mapping.

The mission designers at JPL created a tool to construct 'skeleton' tours – only the Titan flyby parts. It allowed members of the radar team to try to design their own tours, using an Excel spreadsheet. As a sort of thought experiment, one member put together a tour that ignored absolutely every other consideration except maximising the number of Titan flybys – 66. As it turns out, this would have been a difficult tour to fly, with too much activity since each flyby needs several tracking passes to accurately measure the orbit and a couple of burns to trim the orbit into the desired one. Most of the Titan passes flew over the same part of Titan anyway – not ideal for mapping its surface.

From this exercise and the reactions to the A1 to F9 set, the planners, John Smith and Aron Wolf, got a feeling for what was good and what was bad about each. And so in 1996 the teams plunged into a new set of tours, T1 to T18. These were somewhat more refined, in that they included encounters with what are referred to for convenience as 'the icy satellites', in other words the larger moons of Saturn, excluding Titan, which is of course icy but in a category of its own. It involved much more computation to fine-tune these encounters, since flying close to an icy satellite on a trajectory that also meets two Titan encounters requires careful timing.

Another JPLer, Nicole Rappaport, had been busy creating a program called EVENTS, that would run through a tour and find events like flybys, particular viewing geometries, and so on. This gave the scientists something with which to see what a given tour did for them. One of its products was a plot of the groundtrack of *Cassini* over Titan for each flyby, more or less showing which regions would be imaged by the side-looking radar.

Of course, different instruments, and even different investigators on the same instrument team, have very different requirements, which are often mutually exclusive. For an icy satellite flyby, geologists might want a close flyby – but not so close that images would be smeared: the important thing would be to have good illumination. On the other hand, geophysicists interested in the icy satellite's internal structure would want to measure its gravity field by tracking the spacecraft. The gravity effects would be stronger the closer the flyby, the only other consideration being that the flyby should take place when a ground station would have an uninterrupted view of the spacecraft for a couple of hours around closest approach. Or what about the rings? Some measurements like the dust abundance, or the flux of radiation on the rings, require being there (i.e. being close to the rings themselves, at low Saturn latitudes). On the other hand, the big picture, for instance making movies of the spokes spinning around on the rings, requires distant viewing from very high Saturn latitudes.

Innumerable conflicts like these exist and are symptomatic of the breadth of science that *Cassini* is attempting. And yet more considerations come into play. Around each Titan flyby, tracking and manoeuvres must be performed, to line up for the next encounter a few weeks away. In a tour with many Titan encounters, this means that several manoeuvres would have to be performed in a couple of weeks – a very high level of activity. In principle it could be done. JPL controllers have done as much to recover spacecraft from crises before. But they balked at an up-front commitment for month after month. Holidays became another issue. Certainly for something exceptional, controllers would tolerate having to work over Christmas, or the all-important 4th of July holiday – but not four years running. After all, they have families too.

Much unhappiness prevailed – there was an impasse. The scientists felt that ground control bureaucrats were compromising their great science plans because they'd miss out on an Independence Day barbecue; on the other hand, the ground controllers had a budget and felt that it wouldn't stretch to infinity: a line had to be drawn somewhere.

Weaving through this minefield, Smith and Wolf created yet another set of tours: one that was glorious for Science (T-18-3) but would put the ground system through hell; one that sacrificed whatever science had to go to keep within the ground systems' stated – if not universally accepted – rules (T-18-4); and one (T-18-5) that trod a middle path, bending the odd rule here or there if it meant an enhancement to the tour, like a better icy satellite flyby. (Somewhere along the line a T-9-1, T-18-1 and T-18-2 were developed, but fell by the wayside.)

Better tools were coming along, notably one playfully called CASPER (Cassini Sequence PlannER), which could generate perspective views of the saturnian system as seen from *Cassini* and overlay the footprint of the camera and other instruments – to see what could be seen, so to speak. The radar team used a tool called SCAN, which though ungainly, was powerful.

Armed with these tools and a growing realisation of how much

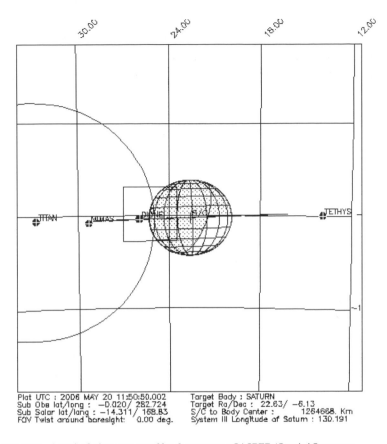

Plot UTC : 2006 MAY 20 11:50:50.002   Target Body : SATURN
Sub Obs lat/long : −0.020/ 282.724   Target Ra/Dec : 22.83/ −8.13
Sub Solar lat/long : −14.311/ 168.83 S/C to Body Center :   1284668. Km
FOV Twist around boresight: 0.00 deg. System III Longitude of Saturn : 130.191

Figure 6.13. A typical view generated by the program CASPER (Cassini Sequence
PlannER), which creates perspective views of the saturnian system as seen by *Cassini* and
can overlay the fields of view of the camera and other instruments. This particular
example shows Saturn looming over the limb of Titan (the circle in the left half of the
image, with its centre marked) as *Cassini* approaches for an encounter with Titan in May
2006. The grid squares are 6° across. Saturn looks about 10 times larger than the Moon
appears from Earth. The box at the centre shows the field of view of *Cassini*'s wide-angle
camera. The field of view of the narrow-angle camera is one tenth that size. Mimas and
Dione are behind Titan.

work would be involved to plan the observations, the project had to
narrow down the number of tours under consideration. The compromise T-18-5 was selected, together with T-9-1, retained because it
offered so many good flybys of both Titan and the icy satellites. In
early 1999, the decision was made to go forward with only one tour.
T-18-5 it would be. And it is truly a grand tour, with 44 reasonably well
spread Titan flybys, viewing geometries acceptable to ring scientists
and plasma scientists, and a great set of satellite encounters – three
with Enceladus and one each with Dione, Rhea, Hyperion and Iapetus.

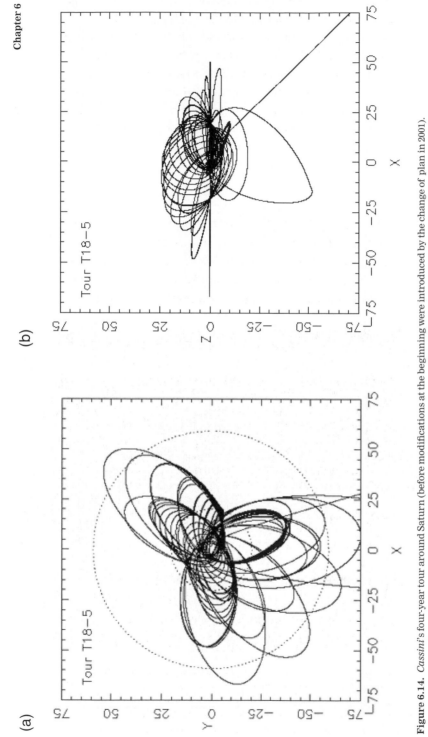

Figure 6.14. *Cassini*'s four-year tour around Saturn (before modifications at the beginning were introduced by the change of plan in 2001).

(a) Looking down over Saturn's pole with the Sun off to the right. The axes are labelled in units of Saturn radii. The inner dashed circle marks Titan's orbit at 20 Saturn radii and the outer dashed circle is the orbit of Iapetus.

(b) Looking into the ring plane. Note the 'cranking over the top' sequence and the many small inclined orbits at the end of the tour:

(In colour, with more details, as Plate 16.)

However, even after the selection, some small changes had to be made. In June of 1999, Smith pulled a rabbit from the hat – a small tweak before the start of the main tour would enable a flyby of Phoebe to within a few thousand kilometres. Phoebe, Saturn's outermost satellite, perversely orbits in the wrong direction, suggesting that it might be a captured asteroid. Previously *Cassini* was to have flown by some 200 000 km away, getting pictures only a few times better than *Voyager*'s grainy blob. The new encounter would really expose Phoebe to close scrutiny.

Visiting Earth

On the 18th of August 1999, *Cassini* whipped past Earth at a cracking pace of 19 km/s. The anti-nuclear protesters tried again to derail *Cassini* in the courts, although the point, now that *Cassini* was in space, was far from clear but Newton's laws of motion won out anyway. When the spacecraft made its closest approach, it was

Figure 6.15. A CCD image taken by amateur astronomer Gordon Garradd from Loomberah, Australia, showing *Cassini* moving against the stars during its flyby of Earth on the 18th of August 1999. The image was taken with a 450-mm (18-inch) Newtonian telescope and a CCD camera provided by The Planetary Society to allow Garradd to search for near Earth asteroids. The 11 dots in a row are images of *Cassini* at 10-min intervals.

a mere 5 km higher than planned at around 1180 km and arrived only 0.6 s late. Earth's gravity swung *Cassini*'s trajectory around by about 20°, corresponding to a valuable boost in speed of some 5.5 km/s.

In fact, several days previously the flyby time was tweaked by 15 s when analysis of the predicted trajectory showed that *Cassini* could pass closer than 25 km to an object catalogued as 7046 – a spent upper stage from an Atlas rocket launched in 1973. The chance of a collision was minute but prudently one of the correction manoeuvres was adjusted to avoid even this tiny risk.

Closest approach was over one of the least populated regions of Earth – the Pacific Ocean. A few observers with telescopes or binoculars saw *Cassini* streaking past and USAF radars also gave the object a glance – good practice for detecting asteroids, perhaps.

Cassini itself was busy. The imaging system took pictures of the Moon, which is a very well-studied object and so a good calibration target for ISS and VIMS. The radio and plasma wave system heard the emissions from Earth – auroral emission and terrestrial radio stations – while the plasma instruments engaged in a massive campaign of measurements together with several Earth-orbiting satellites. *Cassini*'s magnetometer boom deployed and particular attention was devoted to the geotail, the long wake of Earth in the solar wind.

Cassini's radar too was switched on – perhaps the most distant (civilian) radar observations of Earth. Although with the radar, unlike many of *Cassini*'s other instruments, we didn't expect to learn anything about Earth that we didn't know already, it would be the first chance to see what the radar could do. Although the signal would be too weak for imaging, the radar could be operated in a scatterometer mode, giving a profile of radar brightness along the beamtrack. The beam (following where the Sun would be if Earth weren't in the way since the antenna was still acting as a sunshade for the probe) crossed from the Pacific, across the Andes and into Brazil. Not only was there backscatter data but also microwave brightness: just as red-hot objects glow, so cooler objects radiate at longer wavelengths, in the infrared and microwave parts of the spectrum. This microwave brightness, measured with a radiometer, depends on the composition and temperature of the surface.

RALPH'S LOG. AUGUST 1999.

Real data from Cassini! The radiometer data showed the transition
between the mirror-like ocean surface (which basically reflected the
microwave brightness of the cold sky) and the land. A dip soon after
landfall corresponded to the Andes mountains, high and cold. There
was a funny double dip later on in the data. In looking at the data for
the first time just prior to a team meeting, I didn't get time to dig out
comparison data on my workstation but grabbed my pocket atlas on
my way out of the office. On the plane on the way to the radar team
meeting at JPL, I found the plot of the radiometer signal and a map
showing the beamtrack and compared it with the atlas. Sure enough,
the dip corresponded with part of the Brazilian highlands. (I confess I
didn't know about them before then – you learn a new thing every
day!)

Eventually, after comparing the Cassini backscatter and radiome-
ter data with observations by Earth-orbiting satellites and more recent
maps, we were able to establish that this funny dip, which coincided
with a sharp double peak in backscatter, was due to a small reservoir
formed by a fairly recent dam.

Land and sea, the Andes mountains, and a reservoir – and all in
the dark! The radiometer data, which I had not expected to be terribly
interesting, showed great promise. It will be instrumental in resolving
the question of Titan's surface temperature variation with latitude.

Darmstadt, we have a problem

Although catastrophic failures are thankfully rare, they do happen.
More frequent, yet often forgotten, are the so-called 'anomalies' – failures, problems and unforeseen circumstances with consequences that
could ruin a mission but for imagination and hard work on the part of
engineers. Hard work, sadly, interests the media less than catastrophe.
The *Galileo* spacecraft orbiting Jupiter suffered several crippling failures, yet has still returned a wealth of dramatic and important
scientific data. *Mars Global Surveyor*, which in the end pumped hundreds of pictures a day to Earth, nearly broke when atmospheric
changes overstressed its solar panels being used as airbrakes to lower
its orbit. Despite everyone's best efforts, *Huygens* was not to be
immune to problems.

As with many disasters, the first warning signs were vague. At a
Huygens meeting in Padova, Italy, Claudio Solazzo from the *Huygens*
Probe Operations Centre in Darmstadt, Germany, reported the results

of the end-to-end test of the probe relay link during the Earth flyby. As *Cassini* whipped away from Earth, a signal had been transmitted from the ground, pretending to be the probe. Although the signal was received and the Doppler shift measured accurately, the data part of the signal could not be effectively decoded. 'Did he just say they couldn't extract any data?' someone said. 'I think so but I'm not sure,' was the reply. This early announcement was couched in qualification and uncertainty: maybe the test was set up incorrectly. Maybe the nightmare would go away.

No such luck. The test was repeated in summer 2000, with the same result. It seems that the receiver on the orbiter and the transmitter on the probe were not set on the correct frequencies. In an ideal world, the frequencies are the same and everything is easy to check and test. In reality, because the signal from the probe is Doppler-shifted by the orbiter's motion towards it, the transmitted and received frequencies are significantly different. This makes it much harder to test the system. In fact it was not a simple question of the receiver being mistuned. The design flaw is quite subtle: 'inadequate bandwidth on the receiver bit synchroniser'. Better signal strength, lower Doppler shift and the content of the bitstream itself all affect the probability that a given packet of data will be received. Somehow the mistake slipped through the cracks. The Doppler shift of the pure tone of the signal, the 'carrier' (by which the Doppler shift is measured) was tested end-to-end but the data frequencies were not.

RALPH'S LOG. APRIL 2000.

A system administrator here in Arizona, Joe Plassman, came up with a neat Haiku poem that neatly summarised the situation.

> *Cassini listen!*
> *Huygens probe voice very faint*
> *too shrill to detect?*

A Ph.D. student came up with a more unkind version.

> *Huygens probe quiet,*
> *European engineers:*
> *Inexperienced!*

I felt the need to respond, having myself suffered the disappointment a year before of the loss of Mars Polar Lander and the two small pene-

trators it carried, on which I had hoped to measure the hardness of the Martian surface. . . .

Hush, arrogant Yank!
Last year's mission success rate
not exactly great

Exploring Planets
Difficult for everyone
We all try so hard!

The bad news of the test results was somewhat offset by the first pretty pictures coming back from *Cassini* as it began its approach to Jupiter. However, it was a big problem. Although most of the probe data from the last half of the descent would be okay, to lose any data would be a terrible blow. It was yet another repeat of the lesson to spacecraft engineers to test and test again. The loss of the faster, better cheaper *Mars Polar Lander* could have been prevented with a slower schedule and more testing. More testing on the ground could have identified and fixed the *Huygens* radio problem and, if the test had not been done during the Earth flyby, the problem might never have been discovered until the probe reached Titan, when it would have been too late. There were some who had not wanted to perform the Earth flyby test; tests mean people and people mean money, and no-one has money to waste. If the test worked, it would be a little bit of good practice but otherwise would not tell anyone anything they didn't think they knew already; and if it failed, would they know whether the hardware was wrong, or the test was wrong, and what could they do about it? In any case, the hardware was on its way to Jupiter, beyond reach.

Thankfully, prudence prevailed and the test was carried out. Now mission planners would have to use all their ingenuity to come up with a rescue plan. At least there was time to think.

RALPH'S LOG. APRIL 2001.

It is 7.30 a.m. – far too early. I am in my office in Tucson, straining to hear distant voices. Also taking part in the teleconference are some of the DISR engineers downstairs and Jonathan Lunine on his cellphone as he drives in from his home south of Tucson. Five hundred miles to the west, at JPL, engineers and managers are tuned in. On the other side of the Atlantic, a dozen other telephones are connected from almost as many countries. I can't say I like the early hour, but I have been on the other side of these transatlantic teleconferences too. Over in Europe they are hungry, hoping the meeting will end quickly so they can go home for dinner. But there is a job to be done.

It is one of twice-weekly telecons of the Huygens Recovery Task Force, which are being supplemented by lots of homework and several face-to-face meetings. The Task Force is led by German Kai Clausen, who was the Huygens system manager at ESTEC back when I worked there. Like almost everyone else, he is moonlighting from his real job (managing the Integral Gamma Ray telescope project). Also drafted onto the task force are many JPLers, telecommunications specialists from ESTEC and from Denmark, engineers from Alcatel in France and Alenia in Italy, mission designers at ESOC, and many extras from the science teams as required.

How can we fix the link problem? It took over 6 months to develop a solution, during which time we all learned more than we ever wanted to about the guts of the Huygens radio receiver: Viterbi data encoding, temperature drift of bit-synchronisation clocks and so on. Modern digital radio receivers are complicated beasts, and the receiver performance is a complex function of signal-to-noise, bitstream content, and Doppler shift. Because automatic gain controls get switched in, there are even situations where increasing the signal strength makes things worse, prompting suggestions of tricks like blasting radio noise from the ground to fool the receiver into adjusting its automatic settings. One option explored is to pad the datastream with 'zero packets' – like extended coffee breaks in a meeting agenda, they would let the receiver 'catch up'. Of course, zero packets would replace real data but knowing which packets were lost would be better than losing packets at random. A task I pick up, since I straddle the scientific and engineering worlds, is generating synthetic sequences of data, with inserted zero packets, so that the science teams can assess their sensitivity to losses. For example, if every other temperature measurement is lost during part of the descent, maybe that's OK because you can interpolate between the data points you get. On the other hand, some datasets, such as images, straddle many data packets and, if some are lost, it may to too difficult to reconstruct the image from the jigsaw of data that is received correctly. Elsewhere lots of midnight oil is burnt, modeling the performance of the receiver, transmitting tests to Cassini – now beyond the orbit of Jupiter

The only way to solve the problem turned out to be to modify the Cassini orbiter's trajectory. By flying past Titan at a higher altitude – around 65,000 km instead of the original 1500 km, the Doppler shift on the signal will be reduced to a level the receiver can easily cope with. Even so, modifications to the decoding software are required and tricks such as turning the probe on four hours early to warm up the transmitter components come into play to get all the probe data. But there is a price to pay. It can only be done by expending fuel on the orbiter and at first the cost looks unacceptably high, perhaps 150 ms^{-1} of velocity change capacity (or 'Delta-V') – enough to extend Cassini's tour by 3 to 6 years. While the probe scientists like this solution, orbiter scientists are not sure it is a fair compromise.

Jerry Jones at JPL comes up with a solution that looks cheaper on fuel, requiring perhaps only 80–100 ms^{-1} – a big enough drop to make the change palatable. There would still be enough fuel for maybe 2–4 years of extended mission, if nothing else goes wrong and the other major consumable, the hydrazine used to power the thrusters to make rapid turns during high-activity flybys, holds out. There is a catch, though. The landing site has to change so that the probe can communicate with the orbiter as it flies by on the opposite side of Titan from that originally planned. The probe must now land a little north or south of the equator (it is still being optimised as I write) at around 190° W longitude – a little to the southwest of the original landing site.

This extra flyby has to be inserted early in the tour without changing the rest of the tour plan. Cassini will make its first two flybys (Ta and Tb, to avoid disrupting the numbering scheme that has now become familiar to us all) on October 26 and December 13, with the probe still attached. These flybys will also allow good imaging of the landing site, and a confirmation of the winds and atmosphere structure, before the probe is released, nominally on Christmas Day 2004. As the orbiter flies overhead during flyby Tc, the probe will land on Titan on January 14, 2005. The orbiter then rejoins the old T18-5 tour at the fourth flyby, still named T3, in February.

JUNE 2001. OXFORD.

The Cassini Project Science Group (PSG) meeting, usually at JPL but once a year in Europe, is being hosted by co-investigators on the orbiter CIRS instrument in Oxford. They organise a cricket match – a replay of the Cassini versus Huygens soccer match at the launch – but this time the Cassini players win. Brits on both teams smirk at the Americans on both teams fighting the urge to throw away the bat as they are conditioned to do in baseball. Everyone has a good time before getting down to the serious business of the week-long meeting, often running into the evening each day.

> *The main topic of the meeting is the Huygens recovery, together with the tour planning – now extending to all the observations of the rings, satellites, Saturn and so on, as well as the Titan encounters. There is a frightening amount of work to do in the three years remaining before Cassini's arrival at Saturn arrival. Bob Mitchell, the JPL Cassini project manager, always starts his status report with a chart of 'Where is Cassini Now'. For the first time both Cassini and Saturn are on the chart – with Jupiter behind us. We are on the home stretch.*
>
> *The PSG approves the probe recovery plan after some debate. The Surfaces Working Group likes the new tour, as there are some good extra opportunities to view Enceladus during the Ta–c period. Some replanning of the Titan observation programme that had been resolved only after bitter arguments in yet more meetings during the previous year will be needed, but hopefully the changes will not be too large.*
>
> *With the meeting out of the way, I hope to get back to doing some science. Our new HST images show that the north–south asymmetry on Titan has reversed and is back the way it was during the Voyager era (just when you think you understood everything in Chapter 4!). But the asymmetry hasn't reversed at all wavelengths; Titan is doing her best to stay mysterious.*

On the 29th of June 2001, ESA and JPL publicly announced that the recovery plan for the *Huygens* probe hammered out by the scientists and engineers had been approved at the highest level and that the mission teams could go ahead with implementation.

Passing the giant

In December 2000 *Cassini* passed Jupiter. It was a slow, distant flyby but it allowed *Cassini* to observe Jupiter for months, watching the turbulent giant planet swirl. As well as the clouds and bands, *Cassini* observed Jupiter's aurora and listened to the squeaks and howls of the magnetosphere.

Many of *Cassini*'s observations allow the recovery of scientific objectives that had to be sacrificed when *Galileo*'s antenna problem became apparent in the early 1990s. The problem, which throttled *Galileo*'s telemetry to a mere 40 bits/s or so, meant that data-hungry observations like movies of Jupiter's atmosphere had to be dropped. Overall, *Cassini* took more images at Jupiter than all previous missions combined. Other experiments include mapping the microwave emission from Jupiter's magnetosphere (an experiment involving simultaneous observations from ground-based radio telescopes sending their data to school classrooms), and a unique experiment correlating dust measurements on *Cassini* and *Galileo* to determine the speed of dust particles accelerated away from Io by Jupiter's magnetic field.

Planning these observations took great effort, not least because things were held up by the 'software development deferment' that had sounded so innocent in 1992. A year of hard work on the part of the software engineers and the scientists designing the observations got plans together just in time. An added complication, but a great bonus, was that simultaneous measurements could be made with the elderly *Galileo* in orbit around Jupiter as *Cassini* observed from above.

However, as closest approach on the 30th of December loomed, so did a problem. One of the reaction wheel assemblies used to turn *Cassini* started to need more power. Somehow there was more friction than permitted in its bearings. To do the same turns required to point the instruments with thrusters instead of the RWA costs about 0.2 kg of hydrazine fuel per day. Not much, but over the several weeks of the encounter it would add up and deplete this precious resource, which ought to be saved for observations in the saturnian system. The observing had to stop. Contingency plans had to be developed at short notice – a minimum set of observations that wouldn't require so many turns. Maybe something useful could be learned just by running the instruments as *Cassini*'s pointing drifted slowly around?

Although a precious few days of observations were lost, the sequence was restarted once the reaction wheel problem was fully diagnosed and data started to flood down again. This episode doubtless presages crises in the future, when a mad scramble, and yet more hard work, will be required to get the most out of *Cassini* during its busy years ahead. This was just one encounter. There will be about a hundred more!

So on to Saturn and Titan.

The shape of things to come

While *Cassini–Huygens* speeds on its lonely journey between Jupiter and Saturn, back on Earth research on Titan is not suspended. There will be new observations with the giant Arecibo radio dish and with optical/infrared telescopes, new models and new ideas. But as 2004 draws near, the vicinity of Titan itself will become the focus of attention and in the years beyond 2004 it is without doubt where the action will be. The experience of planning *Cassini*'s Jupiter encounter at the end of 2000, which produced over 23 000 images, has made the awesome job ahead at Saturn all too apparent.

Spinning the monster

In the 'old' days of spaceflight, say for example when *Voyager* encountered Saturn, the scientists decided what they wanted to look at and the engineers at JPL would work out what needed to happen in terms of spacecraft manoeuvres. They would send commands like 'fire X thruster for Y length of time' to the spacecraft, which would dutifully implement them. There was a clear division of labour and a well-defined gradient from the scientists' 'high-level' wishes to the space-craft's 'low-level' commands.

It took around six months to plan a *Voyager* encounter. For *Cassini*, with over 100 encounter events, such a pace would be prohibitively slow. The process has to be much more efficient. This has been achieved in part by making *Cassini* far more intelligent than its predecessors. It has an onboard computer program to calculate its own whereabouts and that of every celestial body of importance to its mission (Saturn, Earth, Titan, etc.). Thus, rather than explicitly working out the required turns or even thruster commands like 'Fire

+X thrusters for 200 milliseconds, wait 110 seconds, then fire −X thrusters for 200 milliseconds', controllers can simply say, 'Point half a degree to the left of Saturn's centre'. If everything always went according to pre-ordained plans, this capability would be of only modest importance. But in reality, plans change and spurious things happen. At a given instant in the tour *Cassini* might be starting from a different point than planned. Under such altered circumstances, low-level commands would have to be changed but with high-level commanding much of the program or 'sequence' can stay the same. What is more, parts of the sequences can be re-used for other encounters, since the logic of doing something like stepping the camera's field-of-view to build a 4×4 mosaic of images around a point on Titan's surface stays the same; only the point of reference changes.

The other part of the story is on the ground. The tools for calculating instrument 'footprints' (where the edges of the camera picture fall on a moon's surface, for example), and the computers to run them, are available to all the scientists. With these tools the scientific teams first develop a command sequence for their observation that are to their liking before sending it through a cycle of operations checks.

To an extent this means more work for the scientists. But at the same time, they get to drive a billion dollar spacecraft and avoid surprises in their data. Since they will know how their observation – perhaps a radar scan across Titan, or a mosaic of camera pictures – has been constructed, they know exactly what happened when their data was taken.

Tools don't just mean computer programs. Many of the details of implementing observations, particularly switching from one observation to another, are often thrashed out in real time in what can be contentious meetings where there is neither the time nor the facility to run computer simulations. Under this kind of pressure it is hard to keep in mind what the effects will be on all other instruments if one particular instrument is pointed, say, at Titan. For a while JPL *Voyager* veteran Candice Hansen used a cardboard disk labelled with the spacecraft's axes +X, −Y with a pencil pushed through it to represent the −Z axis – the boresight of the high-gain antenna. *Cassini*'s Project Scientist, Dennis Matson, had the bright idea of printing real coffee mugs with the axes, the fields of view of the instruments, and so on. There were so many instruments that the mugs had to be quite large. The mugs even came with a flight rule: 'only point −Z to ground when mug is empty'.

Cassini's sheer size makes it very stable, as demonstrated by its sharp images of Jupiter. The downside is that *Cassini* takes several minutes to turn itself around. During a close flyby of a saturnian satellite, the spacecraft has to work hard to keep up and, if it has to turn to point different instruments at a target, crucial minutes may be wasted. Oh for a scan platform!

Lights . . . cameras . . . action!

After the Jupiter observations, *Cassini* went into a 'quiet cruise' mode, with only routine engineering activities going on. Gravitational wave experiments will provide breaks in the monotony in December 2001, December 2002 and January 2004 as *Cassini* looks for the ripples in space-time created by cataclysmic cosmic events. During these 40-day-long observations, the reaction wheels will be used to even out the tiny torque due to the faint pressure of sunlight on the magnetometer boom, since using thrusters would ruin the experiment.

The action really begins in early 2004, six months before *Cassini* arrives at Saturn, when the first observations of the saturnian system are taken. Even a month before arrival, the resolution attainable by *Cassini*'s narrow angle camera in images of Titan will exceed the capabilities of the largest telescopes on Earth and it will be possible to study Titan's surface and atmosphere in new detail.

The propulsion system will be pressurised and a tiny burn performed, in part to target the spacecraft at Saturn and in part to clear the throat of the main engine. Then everything happens quickly. As *Cassini* rumbles towards Saturn at 5 km/s, it begins to feel Saturn's gravity and accelerates. An exciting moment will be the flyby of Phoebe, nominally on the 11th of June 2004, a unique and lucky opportunity to get a close look with a variety of instruments at this 200-km object with an organic red tinge to it.

As *Cassini* plunges into the saturnian gravity well, it will speed up to over 20 km/s, coming in along a line roughly 20° south of Saturn's equator. It will be a tense moment, and an important one for the dust detector investigation, when the spacecraft pops through the ring plane. But the hazards have been analysed carefully. The spacecraft passes through a gap between the F and G rings, which lie outside the rings visible to the naked eye, and makes its Saturn orbit insertion (SOI) manoeuvre. At no time in its entire mission will *Cassini* again

Figure 7.1. One of *Cassini*'s most dramatic moments – Saturn orbit insertion in July 2004 – as depicted by JPL artist David Seal. One of *Cassini*'s main engines will burn for some 90 min as the spacecraft whips over the ring plane. Artwork: NASA/JPL. (In colour as Plate 17.)

approach so close to Saturn. *Cassini* fires one of its engines after aligning its thrust vector through the centre of mass (which changes as fuel is burned up), pulsing its attitude thrusters to keep itself pointed correctly. In fact, during the long SOI burn, which lasts one-and-a-half hours, the spacecraft slowly turns with its engine firing in the optimum direction. An on-board accelerometer will sense when the engine has operated for long enough and turn it off. On the way out from SOI, *Cassini* will get a distant look at Titan's south pole before receding to the far reaches of the saturnian system.

The first few Titan encounters will be for many the highlight of the mission. In the new plan of operation decided upon in 2001 in response to *Huygens'* communication problem (see Chapter 6) the first two flybys are scheduled for the 26th of October and the 13th of December. They will be followed by the release of the probe on the 25th of December for entry into Titan's atmosphere on the 14th of January 2005. The probe data will be sent from California to the European Space Operations Centre in Darmstadt in Germany, where ESA-built

Figure 7.2. An artist's impression of *Cassini* releasing the *Huygens* probe. In reality, Titan will be much farther away when this event takes place. Titan is also shown here too far out of Saturn's ring plane. NASA artwork.

spacecraft are controlled. There, the assembled probe investigators will eagerly begin to analyse their data, the culmination of 15 years' work. Doubtless many of them will not get much sleep the day before, whether they are jet-lagged or not!

These will be an amazing few days of meetings and thinking, of tweaking computer programs to analyse the data and working out the best way of presenting them. After those first few days, most scientists will head back to their home institutions and continue making sense of it all while staying in touch by e-mail. But the data will be released after a year or less for scientists all over the world to use and will doubtless be the focus of work for years to come afterwards.

RALPH'S LOG. 1990–2001.

Meetings, meetings meetings. Meetings and paperwork seem to be what drive space projects, not rocket fuel. As the years have gone by, a few of the transparencies for overhead projectors have been replaced by computer-based presentations and the thick binders of background material have been replaced, at least in part, by documents on laptop computers. But some things, it seems, will always be the same – long meetings, dried-up marker pens, bad coffee, grappling with photocopy machines, the perpetual search for e-mail access. Some meetings can have participation by telephone or occasionally by video-conference links. Many have to take place first thing in the morning, JPL time, so that the Europeans can participate before going home for the evening but most of the big decisions and discussions need face-to-face contact.

In my days as an ESA engineer, this meant business-class travel. As an impecunious Ph.D. student, I would often attend the meetings by train, staying with friends. Nowadays, I'm back in the air but, as a uni-versity scientist, strictly economy class. Many of us on Cassini have become all too familiar with the north Atlantic. I always ask for a window seat on the north side of the plane. A glimpse of the aurora borealis on winter flights to Europe reminds me why the fields-and-par-ticles scientists find any interest in their subject at all and heading west I am sometimes rewarded by clipping the southern tip of Greenland. Is Titan like that rolling expanse of pristine white down there, calving into icebergs? Probably not. Maybe more like the smooth, nearly bare rock of north-east Canada, sloping gently into the sea . . . Or maybe Titan looks more like the desert of the southwest USA I see on frequent trips between Tucson and JPL – flat expanses of sediment between moun-tains, with the occasional canyon, dried-up lake bed and dune field. Perhaps it's all of the above. Titan is a wide world too . . .

The wide wild world

Between 1992 and 2001, the nominal landing site for the *Huygens* Probe had been 18° N, 210° E, but that inevitably had to change when a new date was chosen for the landing. It is likely to be at longitude 168° E and latitude either 10° N or 10° S. After the probe enters the atmos-phere, defined to be at 1270 km altitude, its trajectory will be close to the equator. After it slows down enough to throw out its parachute at 170 km altitude, it drifts (eastwards, we expect – see Chapter 4) to its landing spot.

On Earth,10°N–10°S at 168°E is in Polynesia in the Pacific Ocean. How representative is this of our planet as a whole? Not bad, in as much as two-thirds of Earth is covered by water. Despite the detail that *Huygens* will give us, it will be important to put its local findings in a global context with follow-up observations from orbit.

Figure 7.3. A dramatic Titan landscape as visualised by artist Mark Garlick. The scene includes circular lakes. The lightning and the looming presence of Saturn in the sky may both be rather more subdued on the real Titan but Garlick's rendering is more realistic than most. Copyright Mark Garlick, www.space-art.co.uk. (In colour as Plate 18.)

The *Galileo* probe delivered into Jupiter's atmosphere in late 1995 gave puzzling and in a sense disappointing results because it went down in what was found to be a rather exotic part of the planet – a 'hotspot' region of down-welling, dry air, as if the probe were descending in a Sahara of the sky. So, dramatic cloud layers were not seen by its optical sensors and the only lightning detected was distant. Had it hit a more 'typical' spot, it would have detected much more water vapour and perhaps have been buffeted more by winds.

The descent imager might have more hope of seeing something after landing if the probe lands on a solid surface. A cloud of tholin dust might be thrown up and briefly dim the sunlight falling on the probe. This effect has been observed by several Venus landers. Ralph's own little penetrometer experiment is only useful on a solid surface. On the other hand, a splashdown would be much more evocative and probably more survivable.

Pumping and cranking around Saturn

Even after the first flyby of Titan, which pumps down the energy of *Cassini*'s orbit, *Cassini* still has a long-period orbit that takes it on extended loops into the empty reaches of the outer saturnian system.

To be more scientifically productive, *Cassini* has to reduce its orbital period so as to make smaller, shorter orbits and spend more time near the satellites, the rings and Saturn itself.

One of the first, crucial orders of business is to measure how dense Titan's atmosphere is at very high altitudes, since this dictates how brave *Cassini* has to be to make the close Titan flybys that modify its orbit around Saturn. Existing measurements can't help us here much, since there are likely to be variations with season and with the 11-year solar cycle. Every 11 years the Sun's output of ultraviolet light and X-rays increases, as does the number of sunspots. Solar X-rays are absorbed very high in Earth's atmosphere, making it warmer and causing it to puff up. This puffing up causes satellites in low orbits to suffer more atmospheric drag, which pulls them lower. Similarly, if Titan's upper atmosphere is too dense, then the drag during *Cassini*'s flybys could be too high, turning *Cassini* around when it is supposed to be pointed in a specific direction, for example. The ion and neutral mass spectrometer will measure the density of Titan's atmosphere during these first close flybys at a distance of around 1200 km. Together with the density profile determined from the probe's entry accelerometer, these measurements will determine whether *Cassini* gets the 'all clear' to dip down to 950 km during subsequent flybys.

RALPH'S LOG. JUNE 2000.

At the week-long Cassini Project Science Group meeting in Nantes near the French Atlantic coast, there had been some good news from Earl Maize, the spacecraft system manager who had proudly reported its accurate pointing performance – steady as a rock now that the reaction wheels were engaged in the cruise to Jupiter. As usual, the Titan flybys and how they were to be allocated between different instruments was one of the main topics of discussion. Squeezing the most science out of each flyby would depend on how quickly the spacecraft could turn. Another constraint was how much heating from the Sun could be tolerated by the radiators that keep Cassini's sensitive spectrometers cold. The pointing profile at each encounter might be like a slalom course – get from A to B but without pointing at C.

Earl made an off-hand comment about free-molecular flow heating at Titan closest approach. Maybe that would be another gate to avoid, perhaps a subject for future study. I knew a little about free-molecular flow and the aerodynamics of satellites orbiting Earth. Basically the molecules of gas act like free-flying billiard balls because there are too few to make flowing fluid structures. So I made a quick calculation at the back of the meeting room, since the ongoing discussion depended on this issue. I scribbled a quick vugraph and went to the front of the room. In an earlier (less impromptu) presentation, I had softened the impact of some superficially complex mathematics with a slide printed with the large words 'Don't Panic' from the Hitch-hikers Guide to the Galaxy. As I put up my scribbled heating calculation, JPL scientist Scott Bolton who had been chairing the Titan discussion observed that I did not begin with my 'Don't Panic' slide. He was right. Panic. At 950 km, the heating on a surface exposed to the 'ram' direction would get 10 times greater heating than if it were pointed to the Sun.

The spacecraft itself would be quite safe. That had been checked long before. But the effect of this heating on the instruments, which would make their measurements less accurate and possibly permanently degrade the detectors, hadn't been appreciated. Over the next couple of weeks other teams confirmed my calculation and the ram heating issue has become yet another needle through which Cassini's activities have to thread.

The first few Titan encounters are also important to test out the capabilities of each instrument. For example, it may be that the haze is too thick for the smallest features on Titan to be seen with the camera, or perhaps that the northern hemisphere's haze is too thick but the southern hemisphere's isn't. After the first few measurements, it may be appropriate to redo some of the existing plans, for example to explore newly discovered features of interest.

Between April and September 2005, the tour takes advantage of the fact that the rings and Saturn's equator are appreciably inclined to the Sun as Saturn and Titan come out of northern midwinter. As *Cassini* goes behind the rings, scientists can use its radio signals to probe the fine structure of the rings and extract information on the different sizes of the particles making up the rings. The same process in reverse uses radiation from the Sun and *Cassini*'s spectrometers to study even finer particles.

Once this occultation sequence ends, the next year of the tour swings the orientation of the petal-like orbit around to the nightside, so that the spacecraft can explore the particle and field environment of the 'magnetotail', the region of the solar wind downstream from Saturn. During this phase, *Cassini* is close to Saturn's ring plane and most of its flybys of Titan are over Titan's equator. After the magnetotail observations, from mid-2006 to mid-2007, the petal must be brought

back to Saturn's dayside. This will enable long periods of uninterrupted sunlit viewing to make movies of Saturn's weather. Rotating
the petal by means of equatorial orbits would take too long but JPL
mission planner John Smith devised a neat manoeuvre, 'cranking
over the top', which is both faster and views different Titan latitudes.

In mid-2007, a brief equatorial sequence completes the set of close
flybys of 'the icy satellites'. Arguably the most important of these is
an Iapetus flyby in August 2007, within a few days of *Cassini*'s 36th
flyby of Titan. A lot of effort went into tuning the tour so that the
region of Iapetus that would be visible would be well-illuminated in
order to help elucidate the nature of the process that created the dark
material on its surface.

RALPH'S LOG. 1999.

*The T-18-5 tour is an excellent tour, with lots for everyone. Getting
three Enceladus flybys was great news. Cassini would be able to
observe much of its surface on two of the passes and another would be
devoted to measuring its gravity field. However, when imaging scientists started to assess their potential observations in detail for E3, the
last close flyby, there was a surprise. Cassini was in the right place
over Enceladus, the Sun was in the right place over Enceladus so the
target regions would not be on the nightside. But no-one had thought
to check where Saturn was – and as luck would have it, minutes
before the flyby, Saturn would get between Enceladus and the Sun.
Enceladus would go into eclipse! More meetings and teleconferencing
ensued to see if the tour could be modified. One option would put the
encounter in daylight but would chop a couple of degrees of inclination
at the end of the tour – bad news for the magnetosphere and ring
science teams.*

*I resisted the urge to lobby for devoting that flyby to radar – after
all, we could see in the dark. But this would probably not be a fight we
could win. In the end, some small adjustments were made to improve
matters enough for worthwhile imaging.*

The final year-long phase of the tour cranks the orbital inclination
higher and higher with each Titan flyby, so that *Cassini* can look down
on Saturn's pole and on the ring plane, and explore the particles-and-
fields environment at high magnetic latitudes. As well as affording
occultations of Saturn's high latitudes, this phase also entails many
close Titan flybys, with dense coverage of Titan's north-western quad-

rant. Doubtless small adjustments may be made in the years to come, perhaps major ones. 'No battle plan ever survives contact with the enemy', goes the old military saying. However, this is the status quo at the time of writing, and the overall strategy and balance of coverage between scientific targets is unlikely to change. In all there are around 76 orbits of Saturn, each with a close approach to Saturn itself and the rings. There are dozens of other events: close and distant satellite encounters, passages through satellite wakes or magnetospheric flux tubes, low phase angle viewing opportunities, occultations . . . The list is immense.

And of course there are the Titan encounters. Where the Titan encounters occur depends on what is happening in the tour. During the petal rotation sequence, the flybys are only moderately close (2000–4000 km altitude) and are over the equator. During the inclination raise sequence near the end of the tour, the points of closest approach (about 950 km, assuming the atmosphere is thin enough to permit flybys this close) stack up along the prime meridian. The ground tracks of the last couple of flybys in the nominal tour (T43 and T44) in fact sweep very close to the probe landing site, an opportunity to view 'familiar' territory in a new light. Although *Cassini*'s four-year tour is only the seasonal equivalent of six weeks or so on Earth, there may be detectable seasonal changes on Titan over that period.

'The Mimas Ghost' and other mysteries

So what will the *Cassini–Huygens* mission tell us? Of course there is a long list of questions we hope will be answered but predicting what those answers are likely to be is virtually impossible. That is what exploration is for. Like Christopher Columbus's hoped-for shortcut to the Indies, many of our expectations may prove to be incorrect but, at the same time, as the Americas turned out to be for Columbus, more interesting anyway.

First on the agenda will be the origin of Phoebe. Until a flurry of new moons was discovered in 2000, it was the outermost of Saturn's known satellites. Is Phoebe a captured asteroid? Perhaps its impact history, reflected in its cratered surface, may help us understand its origin. Phoebe has been implicated in another of the saturnian system's puzzles, the immense contrast between the faces of Iapetus, one half of which seems to be covered in a dark, presumably organic material. The shape of the deposit, called Cassini Regio, is not quite hemispherical but more like one of the two pieces of a tennis ball. This suggests an origin connected with its orbit around Saturn. A prevailing theory has attributed the dark area to dust knocked from Phoebe landing on Iapetus, or somehow modifying it by impact.

However, in 2000, Toby Owen and colleagues published a study using ground-based spectra of Iapetus in which they found that the laboratory tholins that best match the spectrum of Iapetus are those containing significant amounts of nitrogen. The audacious hypothesis advanced by Owen is that perhaps this material came not from Phoebe but from Titan. The details of this process have yet to be worked out but, if it is the truth, it is a dramatic surprise. *Cassini* should tell us for sure. High-resolution images will indicate the direction the material came from if mountains or crater rims have created 'shadows' on the icy surface protected from deposition. Spectra will help identify its composition better and radar may help indicate whether the material is in a deep layer or just a thin veneer. The dust detector may catch and directly measure the composition and velocity of the infalling material, giving more clues to its origin.

Perhaps with good measurements of Hyperion's composition we can learn if it is indeed the remnant of a larger body destroyed in a collision and, if so, how much larger the original body was. Studying Titan may indicate how much of all that extra material rained down in a century-long bombardment of Titan's surface.

The other satellites too have mysteries. Is the wispy terrain on Dione really some sort of cryovolcanic deposit? Is Tethys' crack of doom, Ithaca Chasma, a deep canyon that stretches from pole to pole, floored with material that slumped from its sides, or slushy material that welled up from within? Is Enceladus' fresh surface and the local enhancement of the brightness of Saturn's E ring at Enceladus' position an indication of ongoing cryovolcanic activity? All these are exciting questions in their own right and will help put Titan in context.

Close to Saturn, the magnetometer will sniff out the subtle details of Saturn's magnetic field, which was only crudely measured by the *Voyagers*. The field is rather boring compared with those of other planets. Saturn's magnetic pole is aligned with its rotational pole unlike on Earth and Jupiter, for example, where the fields are tilted somewhat to the rotation axis and sweeps around. But that in itself is interesting. What can this close alignment tell us about Saturn's interior?

Is the formation of spokes on the rings associated with energetic charged particles in the magnetosphere? And if so, is it a cause or an effect? Perhaps there is also an increase in micrometeoroid flux. Saturn is a paradise for dynamicists – so many satellites, so many interactions and resonances. For example, the fine F ring is collimated by the gravity of the small satellites Prometheus and Pandora, aptly called 'shepherds'. Two of Saturn's satellites, Janus and Epimetheus, switch orbits with each other every four years in a celestial dance – something *Cassini* may be able to observe. Elegant structures like spiral density waves, bending waves and other exotica are manifested in the rings, which are like a giant, beautiful mathematics laboratory. Interesting in their own right, these phenomena may also help understand the processes involved in protoplanetary disks around young stars and in the formation of planets.

During the 1995 Saturn ring plane crossing, when the reduced light from the rings made satellite studies easier, Hubble images showed Prometheus to be 20° or so out of position – more than can be explained by observational error. Perhaps Prometheus collided with part of the F ring, or went close enough to have its orbit changed, possibly like the hypothetical encounter that may have changed the eccentricity of Titan's orbit. In any case, it is a reminder that the saturnian system is a dynamic place and that the rings are ephemeral, evolving structures.

RALPH'S LOG. 1998.

In 1997 Chris McKay, Jonathan Lunine and myself had played with McKay's model of Titan's atmosphere and simpler versions to see what would happen to Titan as the Sun evolved over the next five billion years or so into a red giant. Initially the program crashed 'MIE PARAMETER OUTSIDE RANGE'. What on Earth did that mean? Somehow the haze particles were too big for the optical scattering functions to work. But they had worked before and I hadn't changed the haze parameters. It turned out that the program was working correctly but that the atmosphere, heated by the stronger Sun, had puffed up so much that the haze particles had longer to fall and grow. And the surface temperature didn't change much. Like a fire retardant cushion, the hazy atmosphere was puffing up as the Sun grew more luminous. But then the Sun would also become red, so the ultraviolet flux that produced the haze would drop and the atmosphere would clear. Without the haze's antigreenhouse effect, the surface would warm dramatically, certainly exceeding the ammonia–water melting point, letting water act on the accumulated tholins. Conditions might be favourable for prebiotic chemistry for half a billion years, with Titan covered in a 'primordial gazpacho' (my cold version of primordial soup) – the length of time that it took life to evolve on Earth.

After this work was published, Discover magazine ran a story on it and had an artist make a picture of Titan, on which I was consulted. Above puddles of molten ice crust, Saturn's edge-on rings loomed in the clear black sky, with a ruddy Sun, swollen to fill the orbit of Mercury, almost the same size as Saturn in Titan's sky. It was a neat picture. But some months later a colleague asked 'Would Saturn still have its rings 5 billion years from now?' Not likely. Crap.

Another dynamical feature is the existence of stable points, where Saturn's gravity and that of a large satellite combine to make a sort of cosmic graveyard. These points are in the satellite's orbit, 60° ahead of and behind it, and are named Lagrangian points L4 and L5 after Joseph Louis Lagrange, the Italian–French mathematician who demonstrated the existence of these stable points when a body is orbiting a more massive one. Ground-based observations close to the time of the *Voyager 1* encounter discovered small (30-km diameter) satellites at the Lagrangian points in the orbit of Saturn's moon Tethys. The satellite in the leading point is named Telesto and in the trailing point is Calypso. How did they get there? Maybe closer study by *Cassini* will help.

When *Voyager 1* flew close to one of the Lagrangian points in Mimas's orbit, it didn't see any satellite. But there was a pronounced dip in the flux of energetic electrons trapped in Saturn's magnetic field. Something was absorbing them – perhaps a cloud of dust, too tenuous to see but enough to absorb the particles. Scientists gave the phenomenon the picturesque term 'The Mimas Ghost': something else to explore.

The end of *Cassini*

June 2008 will be the end of the nominal *Cassini* mission. But what then? Superstition, and the basic logic of not counting one's chickens until they have hatched, argues that one should not consider what happens at the end of *Cassini*'s four-year tour. But planetary scientists are a smart bunch and it has not escaped their notice that most space missions keep going after their advertised 'nominal' lifetimes. There is so much to see around Saturn, with so many awesome instruments, that the words 'extended mission' come up every so often in meetings.

Why do all these missions keep lasting so long? Some scientists accuse the engineers of being over-conservative. There may be a grain of truth in this but it is largely an issue of perception. The scientists would be the first to be upset (at the engineers!) if the mission ended prematurely, so there is a natural tendency to be conservative. If engineers are designing a system to work with 95% reliability over four years and a critical component has an 8% failure rate over that time, obviously one such component will not do. Two are needed, either working in parallel or with a switch that will engage the spare when the primary unit fails. The probability after four years that one unit or both units are continuing to operate is then higher than the 95% minimum. Statistics aside, many systems can limp along somehow if they partially fail – usually with careful nursing from the ground, though that implies the expenditure of time and money. The *Galileo* spacecraft suffered crippling failures of both its antennae and yet its tape recorder still returned tremendous amounts of scientific data.

If *Cassini* survives to 2008 without incident, it will nonetheless be old. Some of its instruments will perhaps be less sensitive, their cooling radiators less efficient due to micrometeoroid damage or deposition of organic gunk, their optics degraded a little. Sensitive detectors will have more dead pixels. The radioisotope generators will have run down a little because the plutonium fuel won't have quite the same oomph as it had originally, and its thermoelectric converters, arthritic with radiation damage, will be a little less efficient. Certainly, *Cassini* will not be able to run as many instruments simultaneously and will not be able to turn quite as quickly as in its sprightly youth.

New missions will be under development and there will be pressure to let *Cassini* go into peaceful retirement or to deliberately set it on a course to destruction. However, if *Cassini* continues to work well, the logic of continuing to operate it will be powerful. There will certainly be plenty for it to do. After years of practice, it will be easier to operate *Cassini* and probably with far fewer people than in its heyday. Furthermore, the idiosyncrasies of its instruments will have been ironed out, their calibrations refined. Like an experienced veteran, *Cassini* and its scientists will be able to spot ever more subtle things.

RALPH'S LOG. 2001.

It has seemed strange to many, and indeed on occasion to myself, that one should work on a project that would not reach fruition for another 15 years. In many ways, though, to devote one's professional life to a project like this is just what crusaders and explorers have done for centuries.

A JPL robotics engineer Bob Balaram came up with a ratio he refers to by my name, the duration of a sensor measurement divided by the wait until its operation. For the SSP penetrometer, it is 50 milliseconds divided by about 13 years, or about one in ten billion. It's probably not quite a record. (High-energy physicists doing particle collisions probably win by quadrillions.) All the same, it's indeed a long wait but worth every millisecond for our first touch of Titan, ever!

It is a sad fact that on missions of this duration, not everyone sees the project through. Many talented and charming people – particularly the engineers – move on to other projects. Some scientists do too, especially students who gain their Ph.D. and then move on to a first job on another topic. The finite human life span also takes its toll. Hamid Hassan, the Huygens project manager, died in early 2001 and John Geake of the SSP team in 1998. Al Seiff passed away in 2000. The loss of Carl Sagan in 1996, and of Jim Pollack in 1994 (a student of Sagan and mentor of McKay and others), were felt by many of the Cassini community. But in a sense Cassini and the new data it sends will serve as a memorial to them.

Even after *Cassini* itself ceases operating, its legacy will live on. Just as archived data acquired by the *Voyager* spacecraft is still yielding new insights, 20 years after it was transmitted, *Cassini*'s much more voluminous data return will keep us thinking for decades afterwards.

In fact, both *Cassini* and *Huygens* carry archives of their own. *Cassini* carries a disk on which are digitally recorded some 616 400 signatures sent in on postcards by people from 81 countries. (The original plan was to use a CD-ROM but, as the deluge of cards grew, JPL planners decided to use the much higher capacity DVD-ROM disks, which were just becoming available at that time.) Signatures came from the entire crew of the movie *Contact* and even from Mary Cassini, an Australian resident descended from Jean-Dominique. Jean-Dominique's signature is on there too, from a May 1690 letter held at the Paris Observatory; Christiaan Huygens' signature, from a July 1684 letter to Cassini, is also on the disk. The disk is fixed onto the

spacecraft between two aluminium plates, to protect it from microme-
teorite impacts – it should last for a very long time. *Huygens* carries a
similar disk.

Although *Cassini* will survey Titan from orbit until at least 2008, it
is not too soon to begin thinking about future missions to Titan. While
the detailed design of systems and experiments will benefit from –
indeed require – the results of *Cassini*, some of the technologies and
devices that will be needed will take years to develop.

In 1996 NASA redefined its planetary exploration programme with
a new framework, or 'Roadmap'. This tries to tie missions to 'themes'
or 'quests'. Many of the missions in the plan are the same as they have
always been. The basic science questions, the measurements needed
to answer them and the platforms needed to deliver the instruments
are the same as before. However, the intellectual framework and
synergy between missions is perhaps more coherent. One of the cam-
paigns is 'PreBiotic Material in the Solar System'. This is essentially
code for exploring Europa and Titan, although Triton, comets and
Mars are clearly relevant. The goal is to understand how widespread
the stuff of life is and what processes control its formation, evolution
and distribution – processes that may have a bearing on our own
origins.

A Europa Orbiter was selected in 1998 to be the first of a new string
of outer solar system missions, though no mission's future is ever
secure in the early stages. If it goes ahead as planned, the orbiter will
map in visible light and at high resolution much more of Europa's
surface than will have been accomplished by the fleeting visits of
Galileo. Additionally, a laser radar (or lidar) and radio tracking from
the ground will better define Europa's gravity field and the crust's
response to the formidable tides raised by Jupiter. This information
should indicate the thickness and strength of Europa's ice shell. An
ice-penetrating radar will, it is hoped, sense any layers in the ice and
detect a liquid layer at the bottom, if there is one near enough the
surface.

For Europa, in the context of the PreBiotic campaign at least, the
next step after an orbiter seems to be some kind of lander to investigate any organic chemistry. A longer-term goal, assuming that
unequivocal evidence for a subsurface ocean of liquid water is found
by the Europa Orbiter, is to penetrate the ice with some kind of probe.
This would use nuclear energy to melt its way through the ice, which
would freeze up behind it. Since it will likely be too difficult to send

radio signals through the several kilometres of ice, the probe would leave behind a string of radio transponders, to relay its data up to the surface.

For Titan, the next steps are not as clear, especially since *Cassini* has not yet returned any data. However, we know that much of Titan's surface will only be mapped at modest resolution since the radar in its highest-resolution mode will only cover a part of the surface and the haze will probably prevent optical instruments from achieving a resolution much better than the radar. The only exception will be the tiny part – about a millionth of Titan's surface – imaged by the camera on the *Huygens* probe. We will get relatively little information on Titan's subsurface, where much of the organic material, and perhaps the most interesting stuff, will be hiding. We probably won't know enough about Titan to be able to decide that there is one and only one place to put a more sophisticated lander.

For all these reasons, a future mission to Titan will probably involve some kind of mobile platform in the lower atmosphere. It will have to be below the haze to perform decent imaging and probably below 14 km altitude so that it doesn't get covered in methane ice. Given the large distances to cover and our continuing ignorance of the surface (which might be sticky and impassable), a surface rover looks like an unlikely proposition. Various other ideas have been advanced – aeroplanes, tilt-rotor aircraft, even boats or submarines! Arthur C. Clarke, in his novel *Imperial Earth*, noted the probable usefulness of vehicles like hovercraft on a surface that could have both solid and liquid parts, although in practice such things are unlikely. Hovercraft are not much use on steep slopes.

One of the simplest ideas is perhaps the oldest: balloons. Balloons for exploring Titan were proposed by the ardent planetary ballooning proponent, Jacques Blamont of France, as early as 1978. Titan's thick atmosphere is relatively easy to float in. Unlike the outer planets, which themselves are mostly hydrogen, Titan's atmosphere has a high molecular weight, so a helium balloon will work well. In a non-flammable atmosphere, a hydrogen balloon is not as problematic as it is on Earth. The choice of hydrogen or helium would probably depend on how easy it is to store the gas on the long trip to Titan and how quickly the gas would diffuse through the balloon envelope.

A balloon is condemned to float in the wind – both a blessing and a curse. On Titan this probably means that a balloon will float from east to west, giving good coverage along a line of latitude but not exploring much north or south of it. Also, for half the time, the balloon will be on Titan's night side. Here imaging is difficult (for obvious reasons) and direct communication with Earth is impossible.

It may be that some kind of heavier-than-air vehicle will be the way to go. An aeroplane would need relatively small wings for its weight on Titan, gaining from both the thick air and the low gravity. (Note that, since buoyancy and weight are both due to gravity, Titan's low gravity does not actually help a balloon much.) An aeroplane could be designed to fly at perhaps 15 m/s, a little faster than the rate at which the subsolar point sweeps across the surface due to Titan's rotation. Thus the plane could be programmed to fly along with the Sun staying in perpetual daylight (except for the seasonal eclipses by Saturn), which would allow it to remain in constant communication with Earth.

Helicopters, or aeroplanes with tilt-rotors, would be the most capable platforms for exploring Titan scientifically. As well as survey- ing large areas like a balloon or aeroplane, they would be able to land at chosen sites of scientific interest. The combination of dense air and

Figure 7.4. An impression of a post-*Cassini* mission to Titan realised by space artist James Garry in collabo- ration with Ralph Lorenz. Compared with a 'rover', a helicopter would offer superior mobility and flexibil- ity while taking advantage of Titan's low gravity and thick atmosphere. The design shown here has contra- rotating rotors for compactness and spindly legs to keep the rotors well clear of the ground when it lands. It is shown inspecting the remains of the *Huygens* probe and using spec- trometers to determine the amount of haze deposited on the probe since its landing. (In colour as Plate 19.)

low gravity helps a helicopter tremendously. Even without optimising the design for Titan, a helicopter would need a factor of 38 less power to hover on Titan than on Earth.

Although aircraft are easy to operate on Titan by terrestrial standards, the power required (either by an aeroplane or a helicopter) is still formidable compared with the power available from a small radioisotope source. There seems little hope of avoiding nuclear power as there is too little sunlight near Titan's surface and chemical energy brought by a vehicle, either as batteries, or rocket fuel, or even oxygen to burn with Titan's methane, would not last long enough to make the mission worthwhile.

The reason radioisotope thermoelectric generators (RTGs) used on spacecraft are so massive for their power is that the conversion of heat to electricity by the thermopiles is very inefficient, typically only 5%. In space, however, the reliability of thermoelectric energy conversion with no moving parts is vital. On a Titan aeroplane, since there are moving parts anyway, some kind of heat engine might be better. Using a Stirling engine with an efficiency of 30% or more, six times as much power can be generated for the same amount of plutonium. This efficient type of engine, named after the Scottish clergyman who devised it, uses a single piston to extract energy from gas heated by some external heat source. You can buy classroom demonstration models that will turn a propeller for half an hour on the energy in a mug of hot water. On the other hand, the safe and mature technology of thermoelectric conversion could still be used on a helicopter since, unlike a plane, it can land and spend a 'rest' period trickle-charging a battery during the Titan night with the steady power from an RTG. Then it could fly for a few hours on battery power to the next site.

Whatever the missions of choice turn out to be, and different scientists and engineers of course have their own favourites, it takes many years for missions to rise to the top of the 'to-do' list in the space agencies and there will probably be no firm decision until the first data from *Cassini–Huygens* comes back. However, ideas are already being discussed and such nitty-gritty details as instruments and techniques for navigating around Titan are already being studied.

New senses for a new world

A high-resolution camera will be a key element of the payload of a future mission to see panoramas of Titan's landscape. *Huygens* will image only a tiny part of the surface and exactly what the camera on the orbiter will see is far from clear, in more ways than one. It may be that features as small as 30 m across will be visible with sufficient contrast or clarity to be useful. This resolution is comparable with the best of the *Viking* pictures of Mars. But with *Cassini*'s limited number of Titan flybys, not all of the surface will be covered at this resolution and, in practice, it may be rather poorer because of the haze. To get to 1-m resolution or better, comparable with the performance of *Mars Global Surveyor*, we have to get beneath the haze.

Optical imaging, and even microwave radar, only scratches the surface. We may well see huge expanses of liquid or sludge on Titan and have no idea what lies beneath, or even how deep it is. There may be some opportunities, though, to probe deeper. If *Huygens* lands in a shallow lake, the SSP sonar may measure its depth. Geological interpretations – for example a calculation of how deep a 10-km-wide crater should be – could also tell us approximately the depth of liquid deposits.

A significant job for a follow-up mission will be to comprehensively catalogue Titan's inventory of organics. One important element of that process might be a sounding radar operating at wavelengths of several tens of metres. These are used regularly in glaciology to map out layers in ice sheets, to discover the base of the sheet and to determine whether the sheet is grounded on rock or is floating on a lake. At long wavelengths, the radar signal is not strongly absorbed and can penetrate many kilometres. Such an instrument, which need not be large or power-hungry, could be flown around Titan across its lakes and profile the depths of the organic deposits everywhere. It could also discover structures underneath the surface. A similar instrument was flown around the Moon on the *Apollo 17* orbiter. It found layers several kilometres down beneath Mare Serenitatis. An instrument like this is to be flown on NASA's Europa orbiter, to try to determine how thick the ice on Europa is and whether a water ocean exists beneath it.

Studying Titan's weather is also likely to be a priority task. The *Huygens* probe lasts only three hours and the *Cassini* orbiter flits past all too briefly, spending only about half an hour below altitudes of 5000 km on each flyby. We will get a much better idea of weather on Titan by using basic instruments for weeks or months. Even the simplest kind of barometer – a pressure sensor that is nowadays included in some wristwatches – could detect pressure changes due to methane storms, or the gravitational tides in the atmosphere.

A magnetometer is another instrument that deserves consideration. Although Titan's sludgy surface seems a strange place on which to advocate particles-and-fields 'space physics' stuff, magnetic measurements may help determine the state of Titan's interior. As the magnetic field in which Titan finds itself changes due to Titan's varying distance from Saturn and fluctuations of the solar wind, the response of Titan's interior (and perhaps any bodies of slushy ammonia-rich ice near the surface) can be probed with a simple instrument of only a few tens of grams. What it lacks in sophistication and sensitivity will be more than made up for by continuity and proximity to its target.

Perhaps the biggest questions for a future mission to address are, 'How far did the chemistry get towards the development of life? Are there amino acids? DNA?' As we described in Chapter 3, the tholin sludge produced by atmospheric photochemistry is only an interesting starting point for more complex prebiotic chemistry that may take place when it is in contact with ephemeral liquid water. The sophistication and sensitivity of the chemical analyses, and the tools to meet the physical challenge of getting frozen samples into the instruments, will need great technical effort. But, spurred in part by the prospect of life or its traces on Mars and Europa, many high-tech exobiology instruments are under development. Their exotic names include the microfluidic chemlab-on-a-chip, Raman spectrometers, chemically specific electrode arrays or 'e-noses', labelled fluorescence experiments, and so on.

Where to go?

One of the big challenges in planning *Cassini* was that we have little idea which parts of Titan are more interesting than others. Exploring different parts is important – low and high latitudes, for example, the bright region, some dark areas and perhaps the boundary between them. Only after *Cassini* has started to acquire better data will we know where the attention of a future mission should be focussed.

One of the most interesting places for a Titan helicopter to visit will be the *Huygens* probe landing site. One reason is that this location will have been studied extensively by all the *Cassini* orbiter's instruments – radar, optical and infrared – as well as by the probe camera, radar altimeter, acoustic sounder and so on. It should be the one place we know best and so the location of choice to check the performance of the helicopter's instruments.

Another reason for choosing this spot is that the *Huygens* landing site will be the best place to detect change on Titan's surface. While features would have to change on a scale of kilometres to be detectable from orbit, much smaller changes (whatever they might be – the erosion of a streambed, or the flow of a sludge glacier for example) will be detectable here, since the probe will have taken images during its descent with a resolution down to a few centimetres. Steady slow processes, like the deposition of tholins or erosion by rainfall, might be detectable because of the 'clean slate' that *Huygens* will provide. Its upper surface, and indeed the parachute, should be pretty clean to begin with. Any deposits on them will have accumulated in the six years or so since the probe's descent.

Pictures of the *Huygens* probe, in whatever state it might be, will be uniquely evocative. The astronauts on *Apollo 12* landed near, and walked over to, the *Surveyor 3* spacecraft, which had landed at the base of a small crater 31 months earlier. They removed some components, including the camera, to evaluate the effect of years of exposure to the lunar environment. The science aside, it will be a fitting pilgrimage for a Titan explorer, human or robotic, to seek out Titan's first visitor. It is tempting to think about a human presence on Titan way in the future. It would be a stimulating and spectacular place to visit, if for no other reason than the prospect of human-powered flight. It would have its grim side, though. A dull red landscape, replete with petrochemicals and carcinogens, would be a depressing and unhealthy place to spend much time.

A well-pressurised dome habitat on the Moon would satisfy would-be aeronauts closer to home. So whatever draws people to Titan will have to be more profound than mere pressure and gravity. The chemicals can be handled by machines – automatic factories set up on the surface, or on balloons or perhaps crawling around on the seabed. In any case, the most promising resource, methane as rocket fuel or more likely as a store of hydrogen for fusion power and propulsion, can be obtained without even surrendering to Titan's gravity. It could be collected by spaceships skimming the atmosphere.

Only the prospect of life – indigenous or otherwise – would really justify a human presence on Titan. For plain old exploration, unmanned probes, experienced vicariously by all of us through what's likely to be formidable virtual reality technology, will probably do.

Planet of explorers

Since the start of the *Cassini* project over 10 years ago, many changes
have swept the world. In particular, the advances in information technology have dramatically increased the ease with which everyone can
participate in the exploration of Titan. The first and most obvious
aspect is the growth of the internet and the world wide web. It is now
not only possible, but easy, for professionals, students at all levels, and
even the idly curious, to obtain images and other data from space missions and ground-based telescopes over the web. Whether for a school
report, a picture for the wall, or a serious study, this easy access gives
the public a direct return for their investment (as taxpayers) in space
exploration. We list where data can be accessed in the appendix to this
book.

The communication afforded by the internet not only liberates the
data from the dungeons of the space agencies and research institutions but also unleashes the otherwise untapped talents of the public.
As an example, the SOHO (Solar and Heliospheric Observatory)
spacecraft, another successful NASA–ESA collaborative project, has
routinely taken images of the Sun, monitoring its seething surface
and corona for energetic events. But every so often, these images show
comets plunging into the Sun. A number of dedicated amateurs have
made it their business to trawl through the archives to find these
comets, characterising their numbers, sizes and orbits. Doubtless
similar opportunities will arise among the thousands of images to be
taken by *Cassini*. Data from the other instruments will also be available.

Although the computing power available to scientists has improved

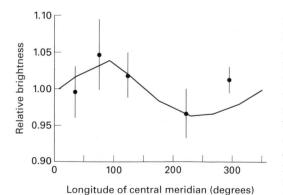

Figure 7.5. The variation of Titan's
brightness with rotation measured at
a wavelength of about 900 nm (the 'I'
band) during the winter of 2000–2001
by amateur astronomer Doug West,
who used an 8-inch telescope. West's
measurements are shown as vertical
bars; he asserts that the variation he
sees exceeds his estimates of the
errors in measurement. The solid
line is a lightcurve from Hubble
Space Telescope data taken at around
the same period of time.

a hundredfold or more over the last decade, some analysis techniques defy automation – the recognition and classification of impact craters, for example. Counting and classifying craters is an important technique for establishing the relative age of a surface, and evaluating processes that have modified it, such as erosion by wind. As the image resolution improves (going from *Voyager* to *Cassini*, say) the number of features to classify goes up, challenging the time available for scientists themselves to do the analysis. However, a trial programme ('Clickworkers' – http://clickworkers.arc.nasa.gov) where volunteers perform an analysis over the internet has shown that such amateur contributions can be valuable in prioritising images for further analysis.

The development of sensitive and inexpensive CCD detectors for video cameras has also brought research-grade instrumentation within the buying power of amateur astronomers. Several amateurs using telescopes as small as 20 cm (8 inches) in diameter and simple spectrometers have been able to take Titan's spectrum, repeating Kuiper's 1944 observation in a few minutes in the backyard.

RALPH'S LOG. FEBRUARY 2001.

I had managed to get a couple of amateurs interested in monitoring Titan but thought I should give it a try myself. I had bought an 8-inch telescope some months before (and had used a video camera to image Mir and the International Space Station) so all I needed was a CCD camera, which costs about $1000, and a quality diffraction grating costing about $200. After a little fiddling to set things up, such as getting the right distance between the grating and the camera, I was in business. The grating let through some of the light to form an image in the normal way and dispersed the rest of the light sideways in a streak to form a spectrum. The spectrum of a red giant star like Betelgeuse was visibly longer than those of other stars. I tried Saturn, which was an impressive bright object and, with the right orientation, separate spectra for the planet and its rings were apparent, with dark methane bands only in the planet's spectrum, just as seen by Kuiper. Tiny Titan was almost lost in the glare of its parent but, with a little effort could be pulled out. To turn the light from Titan into a reflectance spectrum will take more work but, as a quick and dirty test, I divided it by the spectrum of Ganymede since Jupiter was in the same part of the sky as Saturn. Out popped a curve that looked very familiar. Somehow I expected this to be harder . . .

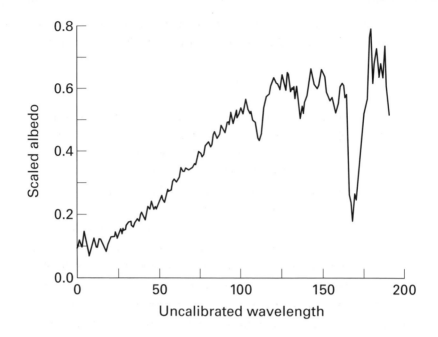

Figure 7.6. Possibilities for amateur spectroscopy. Titan's visible and near infrared spectrum obtained by Ralph Lorenz in his Tucson backyard on the 18th of February 2001. (The curve actually shows Titan's brightness divided by that of Ganymede.) The exposure was for 30 s, with an MX-5 CCD camera and a Rainbow Optics diffraction grating on a 200-mm (8-inch) telescope. The spectrum is probably too noisy for serious scientific analysis but longer exposures, or adding together many observations, will improve the signal-to-noise and perhaps allow amateurs to hunt for clouds on Titan.

While professional study with the latest telescopes and instrumentation will make the headlines, the patient observation of Titan's subtle changes over the years will be a valuable complement. We know already that Titan occasionally brightens when viewed at wavelengths in the wings of the methane windows because of clouds but the only way to know how often or infrequently clouds occur is to keep looking. This is not the sort of project that can compete effectively for time on large telescopes, so small university observatories and dedicated amateurs may be able to make a major contribution. It may be that Titan's weather is strongly season dependent but we'll only know that if the 'Titanwatch' starts now.

The combination of inexpensive instruments and the internet means that people can participate in monitoring Titan even without a telescope of their own. Several robotic observatories are coming on-line. Telescopes in many places, whether driven by full-time

researchers, students or amateurs, either in person or via the int
net, are making it possible to watch Titan closely and continuous.

*   *   *

At a workshop on the exploration of the outer solar system in Houston
in February 2001, Viktor Kerzhanovich, a veteran of the Russian planetary programme now at JPL, presented ideas about airships on
Titan. He called Titan 'the Mars of the outer solar system'. It is an
apposite turn of phrase, not because Titan's size or meteorology is
comparable with that of the Red Planet, but because Titan is likely to
capture the imagination of the public in the same way Mars does.

The *Viking* and *Pathfinder* images from the surface of Mars transformed our perception of the planet: it was no longer a dot in the sky
or the obscure subject of some astronomical study. It was a place; a
place where something might lurk behind that rock; a place where
that rock had evidently been carried from over there by some long-dry
river; a place where dust devils swept past and clouds that formed at
night burned off at dawn; a place that may shed light on the probability of life elsewhere in the universe and on the processes that lead to
its – that is our – creation.

Titan may be all these things and more. It has been, and will continue to be, a great adventure to find out.

247

Appendix. Titan: summary of dynamical and physical data

Radius (surface)	2575 km
Mass	1.346×10^{23} kg $= 0.022$ mass of Earth
Mean density	1880 kg/m^3
Surface temperature	94 K $= -179\,°C$
Surface pressure	1.44 bar
Surface gravity	1.35 m/s^2
Escape velocity	2.65 km/s
Bond albedo[1]	0.29
Magnitude (V_0)[2]	8.3
Mean distance from Saturn	1.223 million km $= 20$ Saturn radii
Mean distance from Sun	9.539 AU[3] $= 1427$ million km
Orbital period around Saturn	15.945 days
Mean orbital velocity	5.58 km/s
Orbital eccentricity	0.029
Inclination of orbit[4]	0.33°
Obliquity[5]	26.7°
Rotation period	15.945 days
Orbital period around Sun	29.458 years

Notes

[1] The bond albedo is the ratio of total reflected light to total incident light.
[2] V_0 is magnitude in visible light at opposition.
[3] AU = astronomical unit. 1 AU = 149 597 870 km, Earth's mean distance from the Sun.
[4] Relative to Saturn's equatorial plane.
[5] The inclination of the equator to the orbital plane.

Bibliography and Internet resources

Books on planetary science in general

Beatty, J. Kelly, Petersen, Carolyn C. and Chaiken, Andrew (eds.), *The New Solar System* (Sky Publishing and Cambridge University Press, 4th edn. 1999). Includes chapter on Titan by Tobias Owen.

Hartmann, William K., *Moons & Planets* (Wadsworth, 4th edn, 1999).

Lewis, John, *Physics and Chemistry of the Solar System* (Academic Press, revised edn. 1997).

Lunine, Jonathan, *Earth: Evolution of a habitable world* (Cambridge University Press, 1999).

Rothery, David, *Satellites of the Outer Planets: Worlds in their own right* (Oxford University Press, 1999).

Stern, S. Alan (ed.), *Our Worlds: The magnetism and thrill of planetary exploration* (Cambridge University Press, 1999).

Weissman, Paul R., McFadden, Lucy-Ann and Johnson, Torrence V. (eds.), *Encyclopedia of the Solar System* (Academic Press, 1999). Includes article on Titan by Athena Coustenis and Ralph Lorenz.

Books on the saturnian system, Titan and the *Cassini–Huygens* mission

Alexander, Arthur F. O'D., *The Planet Saturn: A history of observation, theory and discovery* (Faber and Faber, 1962)

Coustenis, Athena and Taylor, Fred, *Titan: The Earth-like Moon* (World Scientific, 1999). Includes comprehensive bibliography on Titan up to 1998.

Gehrels, Tom and Matthews, Mildred S. (eds.), Saturn (University of Arizona Press, 1984). Includes chapter on Titan by Donald Hunten and others.

Spilker, Linda J. (ed.) *Passage to a Ringed World: The Cassini–Huygens Mission to Saturn and Titan*, NASA SP-533, 1997.

Wilson, Andrew (ed.) *Huygens: Science, payload and mission*, ESA SP-1177, 1997.

Books about managing space missions

Calder, Nigel, *Giotto to the Comets* (Presswork, 1992).

Shirley, Donna, *Managing Martians* (Broadway Books, 1999).

Science fiction

Baxter, Stephen, *Titan* (Harper, 1998).

Clarke, Arthur C., *Imperial Earth* (Victor Gollancz, 1975).

Selected non-technical articles

Lorenz, Ralph 'Lifting Titan's veil', *New Scientist*, 12 April 1997, pp. 34–37.

Lorenz, Ralph 'Did Comas Solà discover Titan's atmosphere?', *Astronomy & Geophysics*, vol. **38** (June/July) 1997, pp. 16–18.

Zubrin, Robert, 'The case for Titan', *Ad Astra*, June 1991, pp. 26–30.

Selected technical publications and articles

Blamont, J. A. 'A method of exploration of the atmosphere of Titan', in D. M. Hunten and D. Morrison (eds.), *The Saturn System* (NASA CP-2068, 1978).

Coustenis, A. and Bézard, B., 'Titan's Atmosphere from Voyager Infrared Observations IV. Latitudinal Variations of Temperature and Composition', *Icarus*, vol. **115**, pp. 126–140, 1995.

Friedlander, A. L., 'Titan buoyant station', *Journal of the British Interplanetary Society*, vol. **37**, pp. 381–387, 1984.

Griffith, C. A., Hall, J. L. and Geballe, T. R, 'Detection of Daily Clouds on Titan', *Science*, vol. **290**, pp. 509–513, 2000.

Hubbard, W. B. and others, 'Results for Titan's atmosphere from its occultation of 28 Sagittarii', *Nature*, vol. **342**, pp. 353–355, 1990.

Hunten, D. M., *The Atmosphere of Titan* (NASA SP-340, 1974).

Kuiper, G. P., 'Titan – a satellite with an atmosphere', *Astrophysical Journal*, vol. **100**, pp. 378–383, 1944.

Lemmon, M. T., Karkoshka, E. and Tomasko, M. 'Titan's rotation: surface feature observed', *Icarus*, vol. **103**, pp. 329–332, 1993.

Lorenz, R. D., 'The life, death and afterlife of a raindrop on Titan', *Planetary and Space Science*, vol. **41**, pp. 647–655, 1993.

Lorenz, R. D., 'Huygens Probe Impact Dynamics', *ESA Journal*, vol.**18**, pp. 93–117, 1994.

Lorenz, R. D., Lunine, J. I., Withers, P. G. and McKay, C. P., 'Titan, Mars and Earth: Entropy Production by Latitudinal Heat Flow', *Geophysical Research Letters*, vol. **28**, 415–418, 2001

Lorenz, R. D., 'Post-Cassini Exploration of Titan: Science Rationale and Mission Concepts', *Journal of the British Interplanetary Society*, vol. **53**, pp. 218–234, 2000.

Lunine, J. I., Stevenson, D. J. and Yung, Y. L., 'Ethane ocean on Titan', *Science*, vol. **222**, pp. 1229–1230, 1983.

Lunine, J. I. and Stevenson, D. J., 'Clathrate and ammonia hydrates at high pressure – Application to the origin of methane on Titan', *Icarus*, vol. **70**, pp. 61–77, 1987.

McKay, C. P., Pollack, J. B. and Courtin, R., 'The Greenhouse and
    Antigreenhouse Effects on Titan', *Science*, vol. **253**, pp. 1118–1121, 1991.

Muhleman, D. O., Grossman, A. W., Butler, B. J. and Slade, M. A., 'Radar
    reflectivity of Titan', *Science*, vol. **248**, pp. 975–980, 1990.

Sagan, C. and Dermot, S. F., 'The tide in the seas of Titan', *Nature*, vol. **300**, pp.
    731–733, 1982.

Smith, P. H., Lemmon, M. T., Lorenz, R. D., Sromovsky, L. A., Caldwell, J. J. and
    Allison, M. D., 'Titan's Surface, Revealed by HST Imaging', *Icarus*, vol. **119**,
    336–349, 1996.

Smith, B. A. and others, 'Encounter with Saturn: Voyager 1 imaging science
    results', *Science*, vol. **212**, pp. 163–191, 1981. (This issue of Science contains
    the initial reports from all the Voyager experiments.)

Thompson, W. R. and Sagan, C., 'Organic chemistry on Titan – surface interac-
    tions', in *Symposium on Titan, Toulouse, September 1991* (ESA SP-338 pp.
    167–176, 1992).

Internet resources

This list of web sites was compiled in mid-2001.

Ralph Lorenz's home page
    www.lpl.arizona.edu/~rlorenz

Bill Arnett's acclaimed 'Nine Planets'
    www.nineplanets.org

NASA home page
    www.nasa.gov

ESA home page
    www.esa.int

JPL Cassini pages
    saturn.jpl.nasa.gov

ESA Huygens pages
    www.sci.esa.int/huygens

Images from the Cassini imaging system can be seen at
    ciclops.lpl.arizona.edu

A Space Library, including a solar system simulator, space art, maps etc.
    samadhi.jpl.nasa.gov

Space artists whose work is reproduced in this book
    Mark Garlick: www.space-art.co.uk
    James Garry: www.fastlight.demon.co.uk

General space art
    www.hobbyspace.com/Art/

Planetary images and data
    pds.jpl.nasa.gov/planets/
    photojournal.jpl.nasa.gov

Hubble Space Telescope
    www.stsci.edu
Links to space science missions
    spacescience.nasa.gov/missions/
Comprehensive listing of astronomy and space links
    cdsweb.u-strasbg.fr/astroweb/
Abstracts of technical papers
    adsabs.harvard.edu

Index